Our Mistreated World

Our Mistreated World

Case Histories of Man's Pillaging of Nature

Edited by John Barnett

DOW JONES BOOKS
Princeton, New Jersey

Manufactured in the United States of America

Standard Book Number: 87128-464-2

Library of Congress card number: 70-138560

Introduction

This is not a cheerful book.

It is a collection of articles, originally printed in The Wall Street Journal, about how we have mistreated our physical world—how we have polluted its air and its rivers, littered its landscapes with junked cars and coal-mine waste heaps, slaughtered its wild creatures and threatened to destroy even ourselves with poisons of our own making.

Much of the degradation of the environment is only too obvious. The most casual observer cannot help but notice the dead fish that wash ashore on Lake Erie, the smog that hovers over Los Angeles and New York City, the black clots of crude-oil sludge that wash onto our ocean beaches after an oil well blowout or a tanker disaster, the mats of rotting waste fibers from paper plants that float down the Hudson River.

But these evidences of gross pollution are only part of the story. Much of the damage wrought on our environment is far less obvious, and both its causes and its total effect are far more complex than the outward signs might indicate. The articles in this book represent an attempt to examine some of these problems in detail. For many readers, they will contain some unpleasant surprises.

There is, for example, the story of air pollution's effect on the weather.

INTRODUCTION

Everyone who has encountered smog knows that it obscures our urban skylines, makes our eyes water and aggravates our respiratory ailments. But scientists also are finding evidence that pollution has begun to alter the weather over hundreds of thousands of square miles of the United States and possibly the global climate as well. Dirty air streaming southward from Chicago and the steel mills of nearby Gary, Ind., brings abnormally heavy rainfall to LaPorte, Ind., some 30 miles away. Scientists say the vapor trails of jet airliners are increasing the cloud cover in heavily traveled air lanes and contributing to a world-wide trend toward cooler temperatures.

Much has been written about the deadly and insidious effects of DDT and other pesticides and creatures that were never intended to be their victims—birds, animals and even people. But there are other, equally deadly and equally insidious poisons in our environment. One is mercury. In sufficient quantity, mercury is a lethal poison for which there is no known antidote. Yet, as one of these articles relates, it has been used as a fungicide applied to many seed grains, often finding its way into the bodies of wild birds that eat the seed in the fields and even into hogs and other farm animals that are sometimes fed leftover seed grain and then slaughtered for human food. Mercury is also used in a number of industrial processes, and it is sometimes dumped into rivers and lakes where it infects fish that later are taken for food.

Not all pollutants are deliberately spilled into waterways or poured out into the air. One chapter of this book describes the little-publicized practice of pumping industrial residues into waste disposal wells — holes sunk thousands of feet into the ground, where porous rock formations soak up the waste like a sponge.

Many of these supposedly "safe" installations are turning out to be polluters after all, contaminating underground water supplies and sometimes even "blowing their tops" to spew thousands of gallons of foul-smelling waste to the surface.

Environmental problems need not always be matters of life and death — at least for humans. Sometimes the question is not whether man will survive, but what sort of world he will survive in. Our abuse of the environment threatens our cultural heritage as well as our natural heritage. Many of the historical treasures of Europe, for instance, are being destroyed by modern civilization. Air pollution is causing ancient buildings and statues to crumble, jet aircraft vibration has cracked stained-glass windows in medieval cathedrals, and tidal disruptions caused by land-fill projects threaten to swamp the entire city of Venice.

And what of the other living things that share the earth with us? Do we want to so dominate our world that we leave no room for many of the wild creatures that have captivated our imaginations in the past? Indeed, putting our own feelings aside, can we deny their right to survive? One such animal is the wolf. Hated by ranchers and farmers and stalked by bounty hunters, the wolf has already disappeared from much of the earth, and naturalists say it is in danger of becoming extinct if it is not protected. Many other creatures are in a similar plight.

The prognosis is not entirely grim, however, and these articles also deal with some of the steps that are being taken to curb the despoilers.

New ground is being broken, for instance, in the realm of environmental litigation in the U.S. Lawsuits brought by conservation groups and some public officials promise to establish the right of the ordinary citizen to breathe clean air and live in unsullied surroundings.

Governmental agencies, meantime, are increasingly conscious of their obligation to curb rather than contribute to environmental damage — a trend illustrated by a new Federal reluctance to ram interstate highways through scenic rural areas and historic urban neighborhoods. Even Detroit's automakers, long regarded by many as the chief villains in the air pollution problem, are hard at work in the search for cleaner ways to propel their cars.

The role we all play as workers and consumers also is discussed in these articles. In large measure, it is our demand for ever greater amounts of energy, ever growing quantities of consumer goods and ever more convenient means of packaging our purchases that has led to our environmental problems. Some economists now are suggesting there is a fundamental contradiction between our faith in an ever expanding economy and our need to halt the despoiling of our world.

Only time will tell, of course, whether the ravaging of our environment can be curbed and whether industrial man can somehow learn to live in harmony with the natural world. But before we can hope to solve our problems we must become familiar with them. It is with the hope of contributing to this knowledge that this book has been compiled.

—JOHN BARNETT
Editor

Contents

Contents

Contents

Our Mistreated World

Air Pollution: There's No Escape

LIKE thousands of other refugees from the Midwest and the East, Joe Sparks Jr., an asthma victim, moved to Arizona from Missouri for clean, dry air, blue skies and bright sunshine.

But year by year he watched Phoenix's air deteriorate until a few months ago the city's skyline disappeared for 11 days as a thick brown pall settled over the city. Nearly all central Arizona, a haven for retirees and persons with respiratory ills, was blanketed by the haze. The smog filled almost every canyon, gulch and gully in the Superstition Mountains. Williams Air Force Base closed its runways.

So Joe Sparks, a 28-year-old attorney in the state's biggest law firm, started a one-man crusade against smog. He helped form a citizens' protest group called GASP (Group Against Smelter Pollution) and founded a scientific-legal organization called Environmental Conscience Inc. to seek evidence against polluters. These days he spends his weekends in airplanes checking on smoke from Arizona's eight copper smelters and taking infrared photographs of smog damage to plants in the desert.

Joe Sparks moved to a clean air haven only to find that there is no escape. Air pollution has now penetrated areas where it was previously unknown—or at least ignored. And its presence is causing citizens like Joe Sparks to react in anger and legislators in many states and counties to pass their first serious air pollution control legislation. Trying to anticipate future problems, the Federal Government has even established air quality control regions in such out-of-the-way vacation spots as the Lake Champlain Valley in Vermont and the Virgin Islands.

But many of America's most famed scenic cities already have serious air pollution problems. In Denver, billed as the mile-high city where on a clear day you can see forever, the nearby Rocky Mountains disappear from view in a brown haze on many days in the winter and spring. A Denver newspaper editor wears a smog mask to drive to work on bad days on the advice of his physician.

In late 1969 Albuquerque started issuing daily air pollution forecasts just like Los Angeles; sometimes residents can't see the Sandia Mountains, only 10 miles away. The leader of a hippie commune who calls himself Ulysses S. Grant campaigned for governor of New Mexico on a platform that would eliminate all taxes except the gasoline tax; he would raise that to a point where it would cover all state expenses and discourage unnecessary car travel, thereby reducing air pollution.

Eugene, Ore., has the dubious distinction of being the home of a new type of smog called "smurk." It's a combination of a murky atmosphere plus smoke from 700,000 tons of straw set ablaze by grass seed farmers in the area from late June through mid-October. Eugene happens to be the grass seed capital of the world. But on Aug. 12, 1969, now known as "Black Tuesday," Gov. Tom McCall called a health and safety emergency when visibility fell to about two blocks in downtown Eugene.

Even Palm Springs, Calif., the winter retreat of millionaires from Hollywood like Frank Sinatra, fears an invasion of Los Angeles smog. Palm Springs city attorney Fred Methine has flown in airplanes to trace "the yellow ribbon of smog creeping down the San Bernardino Freeway," which leads from Los Angeles 100 miles away to the desert resort.

Air pollution monitoring equipment is being set up to determine if pollutant levels are rising in Palm Springs as air pollution officials suspect. Mr. Methine has spearheaded the formation of a regional air pollution association among cities in the area to prevent the establishment of new oil refineries and other pollution-production industry in the Coachella Valley.

The infamous Los Angeles smog has already spread so far that it has infected about 1.3 million stately Ponderosa pines in the San Bernardino National Forest near Lake Arrowhead, 83 miles east of Los Angeles. The U.S. Forest Service has already allowed lumberjacks to log 1,010 acres of smog-damaged Ponderosa pines. Some of the dying giant trees are hundreds of years old. Significant evidence of smog damage to trees has been found elsewhere in California—above the San Francisco Bay area, at the entrance to Sequoia National Park and at Point Loma above San Diego.

Estimates of crop damage from California's smog vary from $30 million to $100 million a year. Lettuce and other leafy vegetables can no longer be grown successfully in Los Angeles County. Vineyards in smoggy Cucamonga, east of Los Angeles, produce only half as many grapes as vines growing in clean air. And in smog-bound Ontario, also east of Los Angeles, orange trees produce only half as much fruit as those growing in clean air.

The smog is now so pervasive in California that when asked where Los Angeles respiratory victims should go to escape, a leading pulmonary disease expert, Dr. Oscar Balchum, says, "I can think of only two places in the state—the area just north of Santa Barbara and Rancho Santa Fe near San Diego." San Diego itself, however, is not free of smog. The city's public health director has called several smog alerts.

As for San Francisco, smog concentrations in the Livermore Valley, a rapidly developing area east of the city, approach those of Los Angeles. And the San Joaquin Valley, the rich farming area in central California, is afflicted with smoky haze—"smaze"—during much of the fall due to agricultural burning.

Dr. Balchum, director of the pulmonary and respiratory disease section at Los Angeles County Hospital, says that most of his respiratory patients are "ordinary working people who can't afford to move out of Los Angeles." So he advises them to buy central air-conditioning or air filters for their homes.

Although power plants, agricultural burning, copper smelters and other industrial pollution are all part of the problem, air pollution experts blame the automobile as the chief contributor to the spread of air pollution.

Denver is a case in point. The city's weather conditions are more conducive to smog than those of any other U.S. city except Los Angeles. Denver is situated in a valley surrounded on all sides by higher ground. As the Chamber of Commerce brags, Denver has sunshine almost every day—an essential ingredient for photo-chemical smog. It also often has temperature inversions during the winter months. An inversion occurs when colder air off the Rocky Mountains blows into the valley and is trapped in a pocket below warmer air and cannot rise. Denver's pollution is trapped in the pocket of cold air.

Since World War II, Denver's population has swollen with ski buffs, mountain lovers and fresh air fans to about 1.1 million. Because the people like to get into the mountains, the city has one of the highest ratios of cars per capita of any city in the U.S.

And Denver has one other problem: The city's altitude causes the internal combustion engine to work less efficiently, thus pumping more pollutants into the air. A gallon of gasoline burned in Denver produces up to 60% more carbon monoxide than the same engine in Los Angeles, for example. It also produces 30% more hydrocarbons in Denver.

All the Denverites' driving, combined with the reduced efficiency of the internal combustion engine, creates smog on a sunny day when the pollutants are trapped near the earth by an inversion. State air pollution officials also say that power plants and industrial firms add more smog.

The result is that Denver's smog problem has been growing worse every year. In January when Governor John Love was about to give his State of the State message to the legislature, Denver became enveloped in a brown haze. Visibility fell to a block. "It was awful," says a state air pollution official. "You couldn't even see the buildings." The governor devoted much of his message to the state's increasing air pollution problem.

Mrs. Ruth Weiner, a pretty young mother of four and a chemistry professor at a women's college in Denver, is among those who have taken up the battle against pollution. She spends many of her waking hours riding around Denver on her bicycle (she refuses to drive a car because of pollution), passing out press releases, collaring state legislators and rounding up scientific and legal experts to testify in behalf of pollution control.

Aroused by the efforts of Mrs. Weiner and other members of her group, Colorado Citizens for Clean Air, the Colorado legislature passed a law establishing a nine-man commission appointed by the governor to set pollution standards. Prior to the law the legislature set the standards. The new law permits the state to fine polluters up to $2,500 a day for an offense—a sharp contrast to the previous law, which allowed only a $100-a-day fine. The state health department set up an emergency smog alert system.

However, one air pollution control official admits privately that the state still has a "minimal" air pollution control operation. "The people are turned on to the problem, but we don't have the manpower or the apparatus yet to do the job," he says.

In Arizona, the eight copper smelters in the state produce more than 51% of America's copper and—according to state air pollution officials—the vast majority of the state's air pollution. Authorities say the smelters pour nearly 10.5 million pounds of sulphur dioxide into the clear Arizona air every day of the year. Smelter pollution is twice that of motor vehicles and two and a half times that of all pollutants produced in the state's two largest cities—Phoenix and Tucson—officials say.

But prevailing winds blow the copper smelter smoke into Phoenix and Tucson, according to county and state pollution experts. The smelter pollutants combine with the cities' automobile and other industrial pollutants to make a brownish haze that's trapped near the earth during an inversion.

Three of the smelters are run by Phelps Dodge Corp.—at Douglas (80 miles southeast of Tucson), Morenci (120 miles east of Phoenix) and Ajo (80 miles southwest of Phoenix). Two are owned by Magma Copper Co.—at San Manuel (about 80 miles southeast of Phoenix) and at Superior (60 miles east of Phoenix). The other three smelters are operated by Inspiration Copper Co. at Miami (80 miles east of Phoenix) and by Kennecott Copper Corp. and American Smelting & Refining Co., both at Hayden (85 miles southeast of Phoenix).

The copper smelter operators deny they are contributing significantly to air pollution in Phoenix and Tucson. I. G. Pickering, general manager of Kennecott's Ray Mines division in Hayden and a spokesman for the industry, says that air pollution monitoring stations have not detected a high concentration of sulphur dioxide in the air in either city. "The people have been brainwashed," he says. "Every wisp of smog is being blamed on the smelters. That just isn't true. About 85% of Phoenix's air pollution is caused by the automobile."

But Phoenix air sanitation bureau chief Robert Taylor says that sulphur dioxide in the atmosphere readily changes into other forms, including sulphuric acid, and that his bureau has no instruments to measure the presence of sulphuric acid in the air. He also cites reports of pilots who say they have followed the smoke from the smelters into Phoenix.

Kennecott's Mr. Pickering says that some of the copper smelters have already spent millions to clean up their pollution and will complete the job—if they're given a reasonable timetable. Kennecott alone has spent $37 million to build a plant, now in operation, that is designed to absorb 80% to 85% of its sulphur dioxide emission. American Smelting & Refining is currently building a large plant of this type, with completion scheduled for December.

"We agree we should have hard goals to reach—but we don't want impossible goals," says Mr. Pickering.

The Arizona House of Representatives is now considering an antipollution bill that would set up a timetable for the copper smelters to clean up their pollution. But it's not known yet what that timetable would be or what the penalties would be for failing to meet it.

Many Arizona residents—from the retirees in Sun City to long-haired students at the University of Arizona at Tucson—are demanding that the smelters stop polluting the air immediately. "The citizen reaction has been intemperate and hysterical," says one copper-company attorney. Adds another company man: "These deep breathers around here go wild. Besides GASP and Environmental Conscience Inc., other citizens groups include the Phoenix Civitan Club's Operation Smog-Stop and the Anti-Air Pollution Drive, an organization of junior high, high school and college students led by a 13-year-old girl.

—EARL C. GOTTSCHALK JR.

1970

King Coal's Legacy

DEEP in the mined-out hills of Appalachia, the message of Earth Day—the urgent need to save the environment—has a hollow ring. The coal industry has long since left Appalachia dead beyond all reasonable hope of reclamation.

The streams around the crossroads hamlet of Lost Creek, W.Va., for instance, run yellow-red with the acid drainage from mines. Huge bare strip-mine excavations slash across miles of once-green mountains. Vacant mine buildings rot and sag. Rusting, abandoned mining machinery clogs rutted back roads.

Such scenes are to be found in the countryside surrounding hundreds of semirural communities scattered across Pennsylvania, West Virginia, southern Ohio and eastern Kentucky. But they aren't solely a rural phenomenon; the urban area around Pittsburgh, the largest city of the Appalachian region, is honeycombed with mines—both active and inactive—and many problems have resulted.

In the Pittsburgh suburb of West Mifflin, for instance, two hastily abandoned homes sit precariously atop mines that started caving in back in 1968. The foundations of both houses are cracked, and the buildings—each worth about $20,000—are total losses to their owners. State officials say thousands of other homes in western Pennsylvania are "vulnerable" to similar destruction.

In another Pittsburgh suburb, Penn Hills, poisonous carbon monoxide fumes seep from cracks in the earth. They come from an underground mine fire that has been burning since 1942. Local officials are spending some $60,000 to pump into the mine a slurry of water and ashes they hope will extinguish the fire, which has forced abandonment of a home and a gas station.

Throughout the Appalachian coal belt, almost two million acres—an area the size of Rhode Island and Delaware combined—are covered with mountains of waste from coal mines. Besides taking up valuable land, much of it in urban areas, the ugly "culm" heaps are often burning. Some have been smoldering for decades, emitting noxious fumes that constantly blanket the areas around them.

Few if any other industries have had such far-reaching effects on the land, air and water of a huge region as coal has had on Appalachia. The coal industry itself acknowledges damage has been done, though it says it is trying to help remedy the situation.

The costs and complexities of the environmental reclamation job are staggering. Federal pollution-fighters estimate that merely to clean up stream-polluting acid mine drainage might cost up to $10 billion. The costs of filling in old mines, smothering underground fires and removing the mountains of coal waste haven't even been guessed at.

Some observers doubt the job can be done at any price. "We might spend billions and then find out we just can't clean up this mess," says Robert C. Scott, a civil engineer who runs a mine-drainage treatment project for the Federal Water Pollution Control Administration at Norton, W.Va.

The rest of the nation can look at Appalachia's

ecological problems with more than casual interest. Midwestern and Western states hold vast coal reserves that are still largely unexploited—and their day is clearly coming.

Coal production in the U.S. is expected to soar over the next 30 years to 1.1 billion tons annually from 575 million tons in 1969. Appalachia has already yielded 25 billion tons of coal during the past century, and many of its richest and most accessible seams are played out. It's inevitable that reserves elsewhere will be tapped increasingly in years to come, and ecologists warn that without great care to protect the environment large areas of the West and Midwest could wind up as desecrated as much of Appalachia.

Why has the coal industry been permitted to cause so much damage in this region?

Basically, because until recent years the costs of pollution and destruction of natural beauty weren't counted in the price of extracting coal. Coal became the economic keystone of Appalachia as U.S. industry demanded more and more coal to make steel and to fuel electric power plants. Coal-producing states fought over markets, and they were loath to saddle mine operators with "unnecessary" cleanup costs. Legislators were often controlled by the coal companies, and even independent-minded lawmakers could be swayed by threats of mine closings and lost jobs.

With few exceptions, mine operators felt little obligation to clean up after themselves. Most had acquired mineral rights under so-called broad form deeds that allowed practically any amount of surface damage to get to the coal—even though the mine operators usually didn't own the surface land. There was nothing to stop miners from ripping out all the coal they could reach, leaving no "pillars" to support the surface. Often only wood timbers supported the mine roof after the coal was gone.

Now the results are all too obvious. When old timbers rot or burn away, mine roofs tumble into the man-made caverns and the surface heaves and cracks. Paved roads split open. Buildings shift and break apart.

In 1969 the state of Pennsylvania spent $90,000 to pump 11,000 tons of fly ash (trapped from the smoke stacks of coal-burning electric power plants) into abandoned mines under Allegheny Valley General Hospital at Natrona Heights, near Pittsburgh. A wing of the hospital was in danger of collapsing because of mine subsidence.

The Kirwan Heights Elementary School in another Pittsburgh suburb has been closed due to mine subsidence. A $12 million high school is to be built in the same area, but school officials say it will rest on caissons extending clear down to the bedrock beneath what's left of the coal seams.

Abandoned mines are by no means the only troublesome ones. Recent coal-mining operations of Republic Steel Corp. near Pittsburgh have produced big earth cracks north of the city, causing Allegheny County officials to worry about the fate of a $3 million swimming pool in the county's North Park. The mining has now been halted, but it's feared that the cracks will continue to spread and destroy the pool.

No less troublesome than mine subsidence is the pollution of Appalachia's streams with sulphuric acid that is formed in coal mines. The Appalachian Regional Commission, a Federal-state body, reports in a massive study that altogether 10,500 miles of rivers and streams in eight states are polluted by mine acid drainage.

Many of the waterways are so polluted that they

can't support fish and plants—they are simply bereft of normal aquatic life. The Monongahela River, which flows through the coal country of West Virginia and western Pennsylvania to help form the Ohio at Pittsburgh, contains so much acid that it remains unfrozen even in the coldest winters.

The Appalachian Regional Commission says industries, municipalities and other users of water incur added costs of $3.5 million annually due to acid pollution. The water must be treated extensively before it can be used for most industrial purposes or as municipal drinking water. Moreover, the corrosive acid eats at water intakes and other equipment in contact with it. River barges in acid water last only about 25 years—15 years less than in normal use.

In Phillipi, W.Va., residents were outraged when thousands of gallons of acid water broke through the sealed mouth of an old mine near the town in the summer of 1969. The water flowed over lawns, killing the grass and finally flooded the town swimming pool, which had to be closed and drained.

Only in recent years have coal-producing states begun to tighten their laws regulating the industry. Not until 1965 did Pennsylvania include mine acid drainage among the pollutants covered by its clean streams law. In 1966 Ohio passed a law requiring coal companies to fill the huge holes left by strip mining—in which the earth's surface is simply stripped away by giant power shovels to get at the coal—but critics say more than 70 amendments severely weakened it. Until 1964 Kentucky required filling only "where practical." Now, however, it has a strong strip-mining law.

The Federal Government until recently had little to say about environmental problems resulting from coal mining. Critics contend that the U.S. Bureau of

Mines historically has shown more interest in promoting the coal industry than in regulating it.

The bureau does not, for instance, have any budget specifically earmarked for environmental affairs—while about $16 million a year goes to its Office of Coal Research, whose main function is to find new markets for coal. Bureau officials argue that their research sometimes turns up valuable information on environmental problems, though they concede such information is an incidental by-product of their work.

A major obstacle to cleaning up existing coal-mine pollution, officials say, is the difficulty of tracing ownership of the thousands of abandoned mines throughout the Appalachian region.

Mining's heyday is long past in Lost Creek, for instance, and most of the companies that once owned the mines are now defunct. Many small mines are owned by families who haven't the means to clean up the pollution even if their ownership could be traced.

Some conservationists contend, however, that some corporate owners of larger mines are deliberately lying low to avoid being assessed the costs of cleaning up mine acid drainage or other pollution. "If you found gold in these old mines, they would come out of the woodwork," contends Richard J. Vanderhoof, who until recently headed the Federal Water Pollution Control Administration office in Cincinnati. (He later was moved to the Dallas office.)

Wherever responsibility should lie, the problem of cleaning up pollution from abandoned mines has fallen on the Federal Government and the states involved. In recent years a number of research projects have been launched with an eye toward finding better ways to clean up mine acid drainage. Though the coal industry is contributing some money to the effort, most of the research funds come from the public coffers. The Federal Water Pollution Control Administration's budget for mine acid drainage research in fiscal 1970 was $3.3 million.

Most states now require active mines at least to neutralize the acid in their drainage waters. This is usually done by channeling the water into a lagoon or settling basin, where lime is mixed with the water. The alkaline lime neutralizes the acid—but the water that eventually flows into rivers and streams, while no longer acid, is far from clean.

The cost of preventing mine subsidence—often the most pressing coal-related problem for urban areas—also is borne largely by government. Usually, the cost is prohibitive except for the most critical projects such as the hospital near Pittsburgh. It costs at least $1.75 a ton to pump fly-ash slurry or other material into an abandoned mine, and some projects have taken as much as a million tons of slurry.

Most states still have no laws to force operators of active mines to take measures to prevent future subsidence. Pennsylvania now requires that pillars of coal be left in underground mines to support the surface in built-up areas, but miners are free to take all the coal in rural areas. Most other states have no restrictions at all.

Under a 1966 law Pennsylvania makes mine operators liable for damage to certain surface structures from active mines. There's no protection for new buildings built over active mines, but mine operators must pay for any damage to structures that were in place in 1966 and are damaged by mining done after that date.

Pennsylvania also offers state subsidence insurance, covering damage from abandoned mines as well as active ones. But premiums are relatively high compared with other types of insurance, and the maximum coverage is $25,000.

Though the coal industry over the years has resisted efforts to force it to clean up its mess, many coal executives nowadays concede that the problems that have traditionally accompanied mining could often be avoided. "These things have happened, but now we're moving aggressively trying to clean up," says Herb Foster, a spokesman for the National Coal Association, a big trade group.

Coal operators cite increased efforts to reclaim strip-mined land as evidence of their new concern with the environment. The Coal Association says strip miners in West Virginia last year reclaimed more than 17,000 acres while mining fewer than 12,000. Nationwide, more than 100 square miles of stripped land was reclaimed last year. "We're trying to catch up and keep up," says Mr. Foster.

"Reclamation" of strip-mined land usually means filling the great gouges left when coal is removed and seeding of the land with grass and trees. One reason coal companies have increased their reclamation activities lately is that most states now require filling and seeding. Often they require performance bonds before mining can begin and return the money to the coal companies only after a mined-out area has been satisfactorily restored.

—Michael K. Drapkin

1970

Furor Over Furs

MRS. Neil Simon no longer will wear the baby leopard coat her husband, the Broadway playwright, gave her three winters ago. Not that it isn't beautiful, she explains. It's just that she feels guilty because "the baby leopards needed it so much more than I did."

Mrs. Simon is worried about possible extinction of leopards and a number of other wild animals. She's part of a growing conservation movement trying to protect these animals from man's demand for their pelts, feathers, skins or horns. The animal lovers are urging consumers to boycott clothing or decoration made from such materials. And more and more women, wealthy trend-setters and middle-class housewives alike, are doing just that—sometimes taking the store-bought skins from their backs.

That's what Mrs. Simon did after she became concerned about the dwindling number of leopards in the world. She says she would gladly burn her coat, except that this would only add to air pollution. "It may just hang in the closet forever," she says.

Such militancy has some U.S. furriers worried about their own skins. They don't enjoy being picketed by their former customers, as has happened in some cities. They say rare furs account for only a small part of their total business but worry that sales of all kinds may suffer. And they have the uneasy feeling that fur may be becoming a kind of negative status symbol.

Actress Gina Lollobrigida may have had that feeling, too, after her appearance in London some time ago. Her maxicoat made of Bengal tiger skins set the fur flying from Britain to New York's fur district. One expert asserted that Miss Lollobrigida's coat probably contained the coats of 10 Bengals; it's estimated there are only 600 to 1,000 left in the world. But the actress was quoted as insisting that the coat had only three skins ("They were big tigers") and that she wasn't personally to blame ("I didn't kill anything").

Miss Lollobrigida's critics aren't buying that argument. Women who buy such products are responsible for the death of the animals necessary to make them, asserts Carol Taylor, a film editor for the National Audubon Society. "They are killing animals in perpetuating the fashion."

Doris Day agrees. "It's absolute insanity to kill and torture animals so that people in Dallas can wrap themselves in a status symbol," says the star of "That Touch of Mink." "I think fake furs are fabulous," she adds.

Estimates vary on the number of species in danger as well as on how close to extinction some are. The most widely accepted authority is the Red Data Book of the International Union for the Conservation of Nature and Natural Resources, an organization based in Morges, Switzerland. As of early 1970, the book listed 889 endangered species, although not all are in imminent danger of extinction. Several U.S. furriers, including Ben Kahn in New York, have announced they won't accept pelts of any animals in the book.

But some conservationists aren't satisfied. "For some reason, scientists are reluctant to list an animal as endangered unless it's about to take its last cough," says Herbert H. Mills, executive director of the Washington-based World Wildlife Fund. "In our view, for example, all the spotted cats are in danger of being wiped out, yet the Red Data Book lists only a few subspecies."

The skins of quite a few "spots"—the fur trade's name for these spotted cats—still find their way into fur coats, according to a report of the Fur Dressers Union. The latest figures are for 1968, when New York fur dressers prepared the pelts of 6,009 leopards, 7,006 jaguars, 4,000 ocelots, 1,656 cheetahs and 159 tigers.

Statistics like these have inspired the Friends of the Earth, a militant conservation group, to push for a "fashion revolt." The group has organized a drive to collect signatures for a pledge not to buy products made from 15 "critically endangered" species: The leopard, cheetah, jaguar, ocelot, tiger, Spanish ocelot, Asiatic lion, red wolf, sea otter, giant otter, polar bear, alligator, crocodile, sea turtle and vicuna. Among the social lions who have signed the pledge are Mrs. Leonard Bernstein, Mrs. George Plimpton and Huntington Hartford.

Some furriers, angered by such militancy, contend their critics exaggerate the situation. "Contrary to the impressions created by over-zealous conservationists," says a report of the Fur Industry Marketing Institute, "our industry does not favor indiscriminate destruction of animals for the sake of their precious hides. It would be quite unrealistic, at the very least, for us to disregard our future source of raw materials."

The report says the "vast bulk" of furs sold in the U.S. come from species such as mink and chinchilla that are ranch-raised for their pelts. As for charges that killing baby seals (often by clubbing) is cruel and inhumane, the report contends: "Although some of the methods may sound barbaric, veterinarians and humane society witnesses have endorsed them and called them as humane as can be."

A new Federal law, the Endangered Species Act, now makes it illegal to import or ship interstate about

135 endangered species in any form, except for research. That means a woman who buys a vicuna coat in South America may be stopped at customs if she tries to bring it into the U.S. The law, of course, doesn't affect European furriers or foreign game poachers, both regarded by conservationists as major threats. And some question whether the law goes far enough domestically.

Gary Soucie, executive director of Friends of the Earth's New York office, argues that the animals protected by the law are "already too far gone" to be helped much. "The ones which need protection are animals, such as the timber wolf or unlisted species of leopard, which are now in abundant supply but could disappear if commercial exploitation continues," he says.

That kind of talk upsets Harry Goodwin, an official of the Interior Department's Office of Endangered Species. "These pure protectionists are getting to be a nuisance," he declares. "People all over the country have been sending me letters full of maudlin stuff—and they're all beside the point. This law is meant to protect endangered species, not to limit the commercial use of wild animals."

About the only people who see a silver lining in the problem are manufacturers of imitation furs, an expanding market. Allura Fashions Inc., a New York maker of synthetic fur coats, for example, has watched sales climb to more than $14 million annually from $4 million when it began offering fake furs six years ago. "If the conservationists are helping take business from the fur people, more power to them," cheerfully says Max Drier, Allura president.

—LINDSAY MILLER

1970

Troubled Waters

NOT long ago oceanographers aboard the research vessel Chain were collecting surface samples from a lonely expanse of the Atlantic south of Bermuda known as the Sargasso Sea. They had planned to study marine life inhabiting the great quantities of drifting seaweed found in the area.

Instead, the scientists made a disturbing discovery. Their nets quickly became fouled with oil and tar—thick sticky globs up to three inches in diameter. Day after day along a 630-mile stretch they cleaned the nets with solvent only to see them gum up again a few hours later. Finally, they abandoned the project in disgust because they were picking up three times as much oil as seaweed.

It wasn't an isolated incident. "Just in the past few years we're finding we can't sail anywhere in the Atlantic—even a thousand miles from land—without finding oil," says Howard Sanders, senior scientist at the Woods Hole Oceanographic Institution, which operates the Chain.

As the vessel's unhappy voyage suggests, worldwide oil pollution—even diluted by the ocean's vastness—is nearing crisis proportions. Beach-goers in such widely scattered spots as the New Jersey shore, Bermuda, the Riviera and the Red Sea complain of gooey black lumps of jellied oil that frequently wash up on shore. Floating oil spills—almost always of unknown origin—each year kill many thousands of seabirds in

North Atlantic and Mediterranean waters, according to surveys by conservationists. Indeed, scientists believe the growing quantity of oil dumped into the sea is threatening marine life of all sorts—and perhaps man as well. The oil industry itself is exhibiting mounting concern.

Where's all the oil coming from?

Ships that routinely discharge oil wastes at sea are the biggest offenders, pollution control experts agree. Tankers, for example, wash out their cargo tanks with salt water after each load. Not infrequently, the washings—along with a heavy residue of oil—are dumped into the ocean. Moreover, passenger liners and freighters often fill their empty fuel tanks with water for ballasting purposes. This highly contaminated mixture is always pumped overboard before the ships enter port to refuel. And vessels of all types normally discharge oily bilge sludges over the side.

Other major sources of unwanted oil include spills from manufacturing plants, refineries and oil terminals. In Boston Harbor alone, a spill of several tons of oil can be expected every three weeks, according to officials of the Massachusetts Division of Natural Resources. Seepage from offshore drilling rigs and spills from wrecks of oil barges and tankers also add to pollution levels.

In recent years, a few widely publicized disasters—like the grounding of the supertanker Torrey Canyon off Britain and the blowout of a well in the Santa Barbara Channel—have focused public attention on oil spills. Yet, damaging as these occasional catastrophes can be, they're only one part of a far larger problem, the experts say.

"It's the day-to-day stuff that's killing us—the chronic oil pollution that nobody reads about in the headlines," says Lieutenant Commander Paul Sova, a

Coast Guard law enforcement officer in New York. Adds a biologist for the U.S. Fish and Wildlife Service: "A great deal of oil is washing ashore all along our coasts. What's its cumulative effect on our environment? That's what we ought to start worrying about."

Statistics on oil pollution are scarce. The Coast Guard lists 714 major oil spills in U.S. coastal waters in 1968, up from 371 in 1966. No one counts spills on a world-wide basis.

Things are expected to get worse. On one hand, world-wide offshore petroleum production is expanding at a rate of 10% a year—and presumably the inevitable minor spills and seepages will grow correspondingly. So far, major blowouts have been rare. But a Presidential panel set up after the Santa Barbara disaster recently warned that by 1980 the U.S. can expect a major pollution incident from offshore wells every year.

Ocean shipments of oil are also climbing rapidly. Capacity of the world's tanker fleet has doubled since 1960 and is continuing to grow. Many of the new vessels are supertankers. These behemoths, with capacities of 100,000 tons or more, will be hauling half of all marine shipments of oil by 1975, it's estimated. (The biggest supertanker afloat today carries 312,000 tons; by comparison, the Torrey Canyon's capacity was 117,000 tons.)

What's more, the imminent tapping of the vast North Slope oil fields in Alaska is adding greatly to pollution fears, especially among conservationists. Tankers will likely be hauling oil through treacherous icebound waters. Even small spills during transport or drilling operations would be especially damaging to the fragile Arctic environment, since oil tends to persist far longer in cold waters than in warm.

Talk of growing oil pollution is most unsettling to Kenneth Battles, co-owner of the Sea Crest Hotel, a resort in Falmouth, Mass., on Cape Cod. He has already had his fill of the stuff.

Sticky black globs of oil washed up on the Sea Crest's beach three separate times in August 1969 alone, Mr. Battles says. Disgruntled guests had to clean their feet with kerosene—and some cut their visits short. "We're sure the oil came from ships heading into Boston, but there's no way we can prove it," he says.

Topping off Falmouth's summer, a barge ran aground on a nearby shoal in mid-September, spewing diesel oil over the town's shoreline. The spill took a month to clean up (the Sea Crest used bulldozers to remove oil from its beach), and for several days Falmouth smelled like a refinery, Mr. Battles says. "The cape should be a refuge from the pollution problems of the city," he adds angrily. "Why drive all the way from New York to find the same damn thing here?"

The Falmouth spill also caused extensive mortality in some 24 species of fish and killed large numbers of crabs, lobsters and scallops, according to scientists who surveyed the scene.

But more disturbing were the subtle effects on the creatures that survived the spill. Weeks later divers from the Woods Hole laboratory found fish and crabs whose natural instincts were strangely altered. Flounder that appeared outwardly healthy allowed themselves to be handled by the swimmers; ordinarily they would have scooted away. Normally skittish fiddler crabs also seemed to have lost their escape reaction; most boldly held their ground as the divers approached.

Max Blumer, a noted organic chemist at Woods Hole, observes that many marine animals produce minute quantities of chemicals that perform functions essential to maintaining the cycle of life. These chemicals act as attractants during the mating process. They also aid predators in locating their prey and, conversely, give warning to potential victims that they're being stalked by predators. Oil—whether from a single big spill or a buildup of repeated small doses—may well upset these vital, chemically triggered processes, Mr. Blumer theorizes, and thus could have a disastrous effect on the survival of many species, including those that are commercially important.

Dumping of oil in the sea may also be creating a new risk of cancer in man. Some crude oil contain compounds that tend to produce cancer in animals. (Researchers, for example, have already found a high incidence of cancerous tissue in certain types of fish taken from the oily waters of Los Angeles Harbor.) Fish and shellfish that are eaten by man can ingest these oils. Hence, Mr. Blumer and other scientists speculate that chronic oil pollution may be leading to accumulation of cancer-causing agents in human food.

In 1966, alarmed over the growing amount of oil in coastal waters, the Coast Guard began a regular schedule of flights by helicopter and airplanes to search out oil slicks and report polluters. Pilots logged 2,000 hours in such patrols during 1968. The Coast Guard has jurisdiction over all vessels within the three-mile limit and some limited powers beyond that.

One problem faced by the patrols is that ships that deliberately discharge oil often do so at night or during periods of low visibility to avoid detection. To combat this, the Coast Guard is developing an electronic sensing device that will aid pilots in spotting oil slicks even in pitch darkness. If the slick is trailing behind a vessel, presumably the pilot could identify the ship and lodge charges against the owner or master.

Yet even in broad daylight, oil surveillance patrols aren't a cure-all—as is indicated by a recent Coast

Guard helicopter flight over New York Harbor. During this 90-minute patrol, the pilot, Lt. (JG) Ray Wirth, and a reporter who occupied the co-pilot's seat easily spotted six different oil slicks drifting rainbow-hued in the bright sunlight. But, as it turned out, none of the sources of the pollution could be positively identified.

In fact, Lt. Wirth didn't even report four of the slicks. To have done so would have been pointless since they were floating far from any conceivable source. (Had the slicks been large enough to require cleanup, he would have radioed word to his home base.)

The two spills he did report were located near possible sources. One was floating alongside a Liberian-flag tanker moored at Bayonne, N.J. The other was located near a Lever Brothers Co. plant on the west side of the Hudson River. (A Lever Brothers spokesman said later that the plant, which makes detergents, soaps and Spry shortening, has occasionally had trouble with oil leaks but that there was no record of a spill that day. He theorized that the oil may have drifted downstream from some other source.)

However, by the time Coast Guard boats reached the scene of the two spills, the oil had drifted away. Hence, no charges were filed.

"It's very frustrating," the young Coast Guard pilot said after the flight. "There's oil all over, but we can't seem to do much about it."

Even when the Coast Guard has the evidence to take a suspected polluter to court, it isn't clear what Government agency should press the charge. The Interior Department's Federal Water Pollution Control Administration is the official pollution control agency, but its powers are limited by a 1924 law that gives it authority only over spills resulting from "gross negligence"—which is tough to prove. As a result, the Army Corps of

Engineers has had to act under an 1899 law that prohibits the dumping of refuse into navigable waters.

Two bills currently before Congress—one in the House and one in the Senate—put considerably more teeth into water pollution laws. But shipping interests are bitterly opposing one feature of the Senate bill, the stronger of the two measures, which imposes unlimited liability upon shipowners for oil spills due to negligence. They contend it will be impossible to get marine insurance unless the liability has a fixed limit.

There's also a move afoot to strengthen an existing international convention designed to limit oil pollution on the high seas. Currently, the convention allows ships to discharge oil wastes more than 100 miles from land. Proposed amendments, which require ratification by member nations, would prohibit dumping of oil anywhere in amounts greater than 16 gallons per nautical mile. However, it's widely recognized that enforcing the convention is practically an impossible task.

For its part, the oil industry is earmarking considerable sums of research money to come up with better ways to clean up oil spills. The American Petroleum Institute, an industry trade group, has encouraged the formation of "harbor cooperatives" in more than 50 U.S. ports. These are volunteer groups of oil industry concerns that pool resources, purchase equipment and establish contingency plans for the quick recovery of spills in their harbors.

In addition, oil company-owned tanker fleets and the larger independent tanker operators a few years ago voluntarily adopted a new method of washing cargo tanks that's designed to keep much of the oil residue on board. It works like this: The contaminated tank washings are transferred to a slop tank instead of being pumped overboard. Then, when most of the oil has floated to the top after 36 hours or so, the relatively clean salt water is pumped over the side. The sludge that remains is consolidated with the next cargo.

Had this new method, known as the "load-on-top" technique, not been adopted, tank washings would add at least 2.1 million tons of petroleum to the sea each year, Shell Oil estimates. But the method has gained wide acceptance and, in fact, has cut that potential pollution by 80%, Shell maintains.

Others aren't so sure. Rear Adm. Roderick Y. Edwards, chief of the Coast Guard's office of public and international affairs, agrees the load-on-top technique has made an "appreciable contribution" to reducing pollution. But he doubts it's 80% effective.

For one thing, Adm. Edwards says, heavy weather often prevents the oil residue from separating from the seawater. Also, tankers that don't carry the same type of cargo each trip usually can't save residues from one cargo without contaminating the next one. Finally, he says, masters of many small, independently operated tankers continue to pump the oily washings overboard simply because the load-on-top technique is too time-consuming and bothersome. "A lot of people are still doing anything they can get away with," he adds.

Oil companies and chemical makers are marketing a variety of detergents that destroy the cohesiveness of oil, thinning it into tiny particles that can be more easily disposed of by bacteria and other natural forces. Increasingly, however, the value of these chemicals is being challenged.

For example, it's widely agreed that the two million gallons of detergents sprayed on the Torrey Canyon spill killed far more marine life than did the oil itself. Since then, "nontoxic" detergents have been developed. But some scientists contend that even these supposedly harmless chemicals are quite dangerous. Since they're designed to disperse the oil into the sea —and hence get it off the surface—the chemicals in effect force-fed the toxic oil to marine animals that might not otherwise be affected, it's believed.

Relatively small amounts of chemicals were applied to the Santa Barbara spill. When the slick eluded booms and other mechanical devices designed to contain it at sea, it was decided to let the oil float ashore, where it could be scooped up by bulldozers and absorbed with straw. This devastated the beaches for months and cut deeply into the area's tourist business. But, according to researchers who are studying the spill, the strategy worked. The high mortality of marine life that occurred in the Torrey Canyon spill was avoided—with the exception of seabirds, which died in large numbers.

—GLYNN MAPES

1969

The Smell of Money

IT was a full-page ad in Time magazine, and it was beautiful. There was the clear, majestic river flowing through the untouched forests. "It cost us a bundle," the caption read, "but the Clearwater River still runs clear." The ad went on to explain that Potlatch Forests Inc., which operates a pulp and paper mill on the Clearwater at Lewiston, Idaho, had spent millions of dollars to control pollution.

There was only one thing wrong. The photograph of the clear Clearwater was taken more than 50 miles *upstream* from the Potlatch plant.

"I can't defend it very well," says Benton R. Cancell, president of Potlatch. But he has an explanation. "We got trapped by a cliche. In Lewiston, there aren't any trees, and our theme is to run things against a forest background." Critics of the company suggest that perhaps another reason the picture was taken upstream is that downstream the Clearwater isn't so clear. At least a dozen times a year, it churns with a foul-smelling white foam that sometimes spurts high into the air, just a few yards from the intake for Lewiston's water supply. Fishermen say the foam and the smell, both products of the Potlatch plant, are sometimes evident 60 miles downstream.

If the Potlatch plant doesn't do much for Lewiston's water, neither does it do much for the air. Last year the Potlatch smokestacks spewed forth 2.5 million pounds of sulphur gases, 1.8 million pounds of particulate matter and 5.5 billion pounds of water vapor. Workers at the plant must drive their cars through a spray device every day to wash off the flakes of sodium sulphate that will otherwise eat away the paint. The hydrogen sulphide the plant gives off often makes Lewiston, which is in a valley downwind from the plant, smell like rotten eggs. Another pollutant is composed of the same chemicals that skunks emit.

The befouled air and water have afflicted Lewiston ever since the Potlatch plant was built there in 1950. Until recently, nearly everyone quietly accepted the pollution as a fact of Lewiston life, or as the company once put it, "the smell of money." (The plant is Lewiston's main industry.) But now times have changed. After the ad appeared, scores of Lewiston residents started organizing to express their concern about pollution from the plant. At a town meeting in the spring of 1970, 150 angry residents confronted two Potlatch executives and accused them of everything from deception (the ad) to genocide (the pollution).

It's the same in almost every city where there's a pulp or paper mill. Aroused citizens—and some state officials—are questioning and confronting the corporate owners. Citizens in one state have barred a company from building a mill in their town, an almost unheard-of resistance to what would have been an economic boon. As a result, the pulp and paper companies are planning heavy expenditures to try to reduce their pollution—though not nearly as heavy as many citizens and officials are demanding.

The pulp and paper companies are being singled out by pollution officials and townspeople because they are among the worst polluters. The chemical "cooking" process that turns wood into pulp (which is then turned into paper) can ruin both air and streams. For instance, the Ticonderoga, N.Y., mill of International Paper Co., which is soon to be replaced by a new facility, has over several decades dumped so much waste into Lake Champlain that 300 acres of lake bottom are coated with guck in depths of up to 12 feet.

Also the pulp and paper makers are more visible targets for pollution fighters because the plants are usually in otherwise scenic areas where there are ready supplies of timber and water.

Officials of pulp and paper makers admit that pollution is a national problem. "Certainly there's a problem," says Mr. Cancell of Potlatch. "Let's not discount that." But some executives say the peril has been blown out of proportion by students and others who have taken up the environment as a cause. Referring to the controversy over pollution of Lake Erie, Bernard W. Recknagel, executive vice president of St. Regis Paper Co., says: "The idea that Americans have a cesspool for a lake gripes the hell out of me. A fish would have a delightful time in 75% of Lake Erie today."

Executives at nearly every company emphasize that they are planning millions of dollars in outlays to curb pollution. International Paper, whose pollution abatement spending jumped to $15 million in 1970 from $4.5 million in 1969, recently announced that it will spend $101 million to fight pollution over the next four years. (Among other things, the higher outlays by the industry probably will lead to increased prices, some executives say.)

The future plans do little to assuage many residents of mill towns, however. "Potlatch keeps telling us about how much money they have spent on pollution controls," says David Hadley, a college student who organized the campaign against Potlatch in Lewiston. "Well, of course they've done something. If they were operating with no controls at all, we'd be dead."

Mr. Hadley has a point. The pollution from pulp mills is a proven health hazard. Dr. Benjamin Ferris, a professor at Harvard's school of public health, compared the town of Berlin, N.H., which has a pulp mill, to a similar-sized town in British Columbia with no major air pollution. Dr. Ferris found that Berlin residents had twice the rate of severe chronic bronchitis and emphysema.

Industry officials say they are aware of the health hazard, and they insist they would eliminate it if they knew how. Their newer plants have the most modern equipment available, many executives contend.

There is evidence to refute this, however. The most effective pollution-control system for pulp and paper mills was developed about 25 years ago in Sweden. Yet only one new U.S. mill—owned by American Can Co., an industry "outsider"—employs this process, which eliminates the rotten egg smell. (Another mill, an older renovated facility of Crown Zellerbach Corp., has converted to the Swedish process.)

The American Can mill, at Halsey, Ore., emits relatively few pollutants. One company official claims he actually climbed to the top of a smokestack at the mill and stuck his head in the plume—a suicidal act in a conventional mill. Now officials of many companies are talking enthusiastically about the "new" process and planning more mills using it. They say they didn't try the method years ago because it is slightly more expensive.

One reason the executives now are seriously considering the Swedish process is because they are suddenly getting a lot of pressure from state regulators. The states of Washington and Oregon have recently issued strict new controls for pulp and paper mills, for instance. And the state of New Mexico recently turned down a company's request to build a mill in the little town of Algodones.

"We're one of the few areas left in the country that has the option of retaining a quality environment," says Larry J. Gordon, state director of health and social services. "The public recognized this and became aroused." In addition, Mr. Gordon says that "the company could not guarantee it could meet our air pollution regulations, even though it was amazing to note how their technology increased week by week as we held the line on our requirement."

The company, Parsons & Whittemore Inc., then decided to locate the mill in Colorado, but it is running into similar problems there. "At this time, we're not all that interested in trying to fight the battle," says George Boylan, a company vice president.

A politician in Washington state says that people there, too, "are up in arms." Pollution in Washington "has become a big political issue, and there are many political careers revolving around it now," he says. Many residents of the state charge their scenic land is being ruined by the pulp and paper mills, but Robert Stockman, director of the state office of air quality control, is satisfied with the cooperation he is receiving from the industry. He declines to disclose statistics on the amount of pollutants emitted by mills in his state, saying, "As far as I am concerned we are very pleased with the attitude of the industry. I don't intend to irritate them" by releasing the figures.

Floyd Oles, city manager of Tacoma, also speaks out for the industry. "I'm primarily interested in the payrolls and not in the smell," he says. He asserts the agitation against pollution is led by people with "long hair and straggly beards." He doesn't question their motives but he thinks they are playing into the hands of subversives. "The Communists are ready to take over anything in the way of a disruptive effort," he says.

The people across the country who are more interested in the smells than the payrolls say the smells and other pollution from mills are getting worse instead of better. In many instances, they appear to be right. The pollution-control equipment on mills generally works most efficiently when mills are operating at an average rate. But for the past couple of years, with demand for paper and pulp strong, the mills have been operating at such a high rate that the control equipment has become much less efficient.

In Lewiston, for instance, the Potlatch mill's daily output rose 6% between 1967 and 1969. In the same period, however, the emission of sulphur-based gases increased 2.7 times.

This, too, upsets some residents of Lewiston, and they are speaking out. "There was a time when people were afraid to protest pulp mill pollution," says Shirlee Hennigan, a Lewiston teacher. "Potlatch would say, 'If you don't like it, we'll take our plant elsewhere.' Now people aren't afraid anymore. They know there aren't any more places the pulp mill can go."

—STANFORD N. SESSER

1970

Europe's Tarnished Treasures

IT'S another rainy morning in Venice. You arise in a once-lavish, centuries-old ground-floor apartment and curse the weather. Then you raise the bed to the ceiling on pulleys and await the flood.

Such practices are increasingly common for thousands of Venetians. As canal dredging and land filling disrupt the Venice lagoon, floods threaten to reduce the city of Renaissance treasures to a ghost town within decades. Meanwhile, air pollution rots priceless statues and growing tourist mobs litter famous squares. A religious hall with 16th century frescoes has become a basketball court.

Venice's problems underscore an alarming trend: Modern civilization is fast destroying great parts of Europe's heritage and environment. Much of what attracts millions of tourists from America and elsewhere to the continent is losing its original character or vanishing altogether. Soaring growth of industry, auto traffic, new construction and mass tourism all mean unprecedented pressure that is likely to grow, experts say. They're hoping increasing citizen pressure will help stem the tide.

The problem seems to be everywhere. In Paris, parking lots encroach on the broad walks of the famous

Avenue des Champs-Elysees. In London, high-rise office and apartment buildings are replacing graceful Georgian structures. In Athens, air pollution, aircraft vibration and tourists have damaged the Parthenon so much that worried experts are considering spraying the ancient temple with a plastic protective film or barring visitors. And in Ferrara, Italy, the former Church of San Michele has become an auto garage where hydraulic jacks lift autos toward an ornately decorated domed ceiling.

"With many of Europe's great treasures, the damage of the last 20 years has been greater than that of the previous 2,000," says Hubert Howard, an Englishman who is president of Europa Nostra, a growing federation of groups dedicated to saving Europe's treasures.

One example is a Renaissance house that Mr. Howard owns in Rome. "The frescoes had survived everything beautifully since the Renaissance, but just in the last several years the vibrations of traffic have started to knock them off the walls," he says.

It's no wonder that traffic is becoming a problem in Rome and elsewhere. The auto population of the six-nation Common Market has soared to more than 34 million, twice the figure of 1961. Moreover, the trend is gaining momentum; new auto registrations climbed to 6.1 million in 1968 from 1.6 million a decade earlier.

Industry is thriving. Common Market chemical output, for instance, is more than two and one-half times the 1958 level. The Common Market's total output of goods and services soared to $380 billion in 1968 from $160 billion 10 years earlier.

For many Europeans, of course, such economic dynamism means better jobs, higher wages and the wide availability of material goods that were so scarce a few years ago that only the rich could afford them.

But more and more Europeans are beginning to regard their new affluence as a mixed blessing—a situation that's nowhere more evident than in Venice.

Much of the city's flood problem can be traced to two industrial zones—each bigger than Venice itself—that have been built by filling in part of the Venice lagoon. The city is situated within the lagoon on more than 100 islets interconnected by canals. By reducing the marshland that absorbs water flowing through the lagoon, the zones greatly increase flooding, experts say.

Indeed, records show that floods have become markedly more frequent after each zone was built. There have been as many unusually high tides in the past 12 years since the second zone was built as in the previous 90 years. In recent years, the ground-floor apartments of at least 15,000 Venetians have been flooded each winter, often with two feet of water. At times, the beautiful and famous St. Mark's Square becomes a lake.

Some 20,000 irate citizens have petitioned the city to halt work on a third industrial zone in the lagoon, and the city has responded by suspending the project —at least for the present.

But at the same time, canal dredging to accomodate large oil tankers has increased the speed and force of incoming tides, experts say. And while the water rises, the city sinks. It has always sunk 10 or 12 centimeters (four to five inches) a century, but now it's sinking five times that fast as increased industrial and household use of water lowers the underground water table.

The United Nations Educational, Scientific and

Cultural Organization (UNESCO) warns that water could permanently cover the city in 70 years. "If nothing is done, our city is finished, that's all," says Countess Anna-Marie Cicogna, a worried civic leader.

Air pollution may beat the water in destroying some of the city's treasures. Roger Clisby, an American art historian working in Venice points unhappily at a 15th century marble angel. Until five years ago, the historic work had lost merely a finger. Since then, one whole hand, part of an elbow and part of the face have crumbled to dust. "It's like leprosy," says Mr. Clisby.

The same problem is evident throughout Europe. Instead of merely darkening the surface of statues, as many complacently believe, air pollutants combine with calcium carbonate in marble to form calcium sulphate. Because the latter substance takes twice the volume of the former, the expansion breaks the stone.

Such developments are grim news to art-loving tourists. But the fast growing numbers of tourists themselves also are threatening Europe's treasures and environment. With rising European and American incomes spurring travel, 28.6 million foreigners visited Italy alone in 1968; nearly double the figure 10 years earlier. Other countries also report unprecedented tourist booms.

Not all these tourists respect art. Visitors walking around inside the dome of the Santa Maria del Fiore Cathedral in Florence have completely obliterated the bottom four feet of the fresco with graffiti. On a fresco in the Church of Santa Caterina in Galatina, Italy, handwritten messages about who Carlo and Bill love cover the Madonna's blue robe.

Tourists also pose more unusual threats. The French government had to bar tourists from the famous Lascaux caves; the breath of visitors caused a biological growth to form on the caves' attraction, Stone Age paintings. More predictably, perhaps, the British Ministry of Public Works must keep a supply of spare metal fig leaves because people keep stealing them from the nude statue of Achilles in London's Hyde Park.

More tourism requires more transportation—and this leads to more problems. At the cathedral in Troyes, France, jet aircraft vibrations have broken stained glass windows and caused gargoyles to fall. Throughout Europe, squares like Rome's graceful Piazza Colonna now double as parking lots.

In Venice, private motorboats and public transport water "buses" called "vaporetti" have replaced all but 500 of the 10,000 gondolas that once served the city. Waves from the motor boats menacingly wear down the mortar of Venice's many canal-side buildings.

Of course, even the staunchest conservationist sees that motorboats are faster than gondolas. Few deny that factories, cars, airplanes and tourism bring many benefits. But efforts to bring progress sometimes produce exactly the opposite result. To check erosion, Venice tried lining some of its famous canals with concrete. The snag: Venice depends on microbes that thrive in the mud bottoms of its canals to destroy some of the sewage dumped into the water. Thus the canals that smelled the worst were those with the "modern" and "progressive"—but mudless—concrete bottoms. The experiment was stopped.

Often, though, the problem is more one of neglect or abuse than of progress gone awry. On Murano Island in Venice's lagoon, an aristocratic Renaissance-era clubhouse today forms part of a run-down glass factory. Its once-ornate gaming room, whose frescoes depict scenes from Dante, is stacked with bags of sand.

The room once offered a panoramic view of Venice, but now the windows are boarded.

One Sunday a few years ago a congregation at Oriago, Italy, went to Mass and then simply pulled down its 17th century church. The members tossed the church's 300-year-old works of art onto the town dump. They had decided a new church would bring the town more prestige than an old one, says Renato Padoan, superintendent of documents in the Venice region.

In recent years, there has been a growing effort to halt the destruction of natural and man-made wonders. Europa Nostra, for instance, was formed only in 1963, but it now claims nine million members through 37 member organizations throughout Europe. One of these groups is Britain's National Trust, which has 170,000 members.

The National Trust preserves stately houses and other properties. In a recent effort called "Operation Neptune," it raised the equivalent of $3 million and sheltered, either by outright purchase or obtaining perpetual protective covenants from owners, 275 miles of England's often desecrated coastline. Other Europa Nostra organizations pressure governments and other organizations to prevent abuses.

With public pressure rising, governments increasingly respond. West Germany recently started a 10-year, $2.5 billion program to clean the polluted Rhine River. Following a U.S. proposal, the North Atlantic Treaty Organization established a high-level committee to foster an allied fight against environmental problems.

—ROGER RICKLEFS

1969

Ah, Wilderness: Smogbound Yosemite

THEY come by the tens of thousands to scenic Yosemite National Park, enduring hour-long traffic jams at the park gates, searching long after that for a place to park and room enough to set up temporary housekeeping. Finally, they establish themselves in a "tent tenement" more crowded than the cities from which they came.

They shop at supermarkets-in-the-wild, hang laundry over tent ropes, feed sandwiches to the bears, and turn their ankles when they try to go hiking in wedgies and capri pants. They have come to get away from the city—but they have brought the city and its ills with them.

Ride with park ranger David W. Huson on patrol. Turning down a dusty lane, he slows his car as some screaming children scamper across the road. The damp night air, heavy with a pall of eye-watering smoke, is cut by the blare of transistor radios, the clatter of pots and pans, the roar of a motorcycle, and the squeals of teen-agers. Except for hundreds of shiny aluminum trailers and multicolored tents squeezed into camping areas, this might be any city after dark.

Indeed, the spectacularly beautiful park is wrestling with problems that would give any fair-sized city cause for alarm—a soaring crime rate, incredible traffic snarls, juvenile rowdyism, and even smog. All are traceable to a sheer overcrowding that threatens many national parks during the summer vacation season. The crush is sure to get worse in years to come if population growth and the American yen for the great outdoors continue to outstrip the growth in parklands.

Park authorities fear the mounting crush may rob many parks of all their charm before long. To spread the crowds a little thinner, the Federal Government has been creating new recreation areas at a rapid clip. But there still is less recreation area available per citizen than there was 20 years ago.

In the mid-1950s the National Park Service launched a 10-year program to provide existing parks with additional campgrounds, more housing for park employes, and other needed improvements. But the program's planners failed to anticipate the full extent of the tourist explosion. "Visitor use has far exceeded what was contemplated when the program was formulated," says Clarence P. Montgomery, assistant director for administration at Park Service headquarters in Washington.

A number of possible solutions to the crowding problem are being discussed there—including the use of buses to bring people into parks and restrictions on auto traffic. "It's automobiles, not people, that are crowding the parks," says one high official of the Park Service.

But Yosemite, at least, has a people problem too. The park is huge—1,200 square miles—but 90% of the park's two million or so visitors a year head for Yosemite Valley, 10 square miles of meadowland and forest ringed in by 3,000-foot granite walls, waterfalls cascading from their heights.

On an average summer day, the valley will hold slightly more people per square mile than Cook County, which includes Chicago, and three times more per square mile than Los Angeles County. Even at this, the valley is not as crowded as some other national parks; tiny (912-acre) Platt National Park in southern Oklahoma, for example, accommodates about 16,500 people on the average Sunday.

The crowding at Yosemite distresses its rangers and many visitors who complain that campgrounds are becoming tenements. At one of seven camps in the valley patrolled by ranger Huson, trailers and tents were crammed within three or four feet of each other over Memorial Day weekend; parked cars spilled over onto the roadway. "Wall to wall campers," sighs the ranger. (He and his fellow rangers like to joke that if one stake was removed, several adjoining tents would collapse.)

On Friday evening of that holiday weekend, Mr. and Mrs. Arthur Enos of Lafayette, Calif., crawled through bumper-to-bumper traffic for an hour to get into the valley, and spent another hour driving through every campground in a futile search for an empty site. On their second time around, Mrs. Enos walked ahead of the family camping truck, carrying a lantern.

Long after midnight, she finally found a spot "after stumbling over rocks and tents and waking everyone up," she says. Another family of campers, Mr. and Mrs. Anthony Quill of San Francisco, was luckier; they came early and found a campsite—only to awaken in the morning and find themselves walled in by at least half a dozen trailers and as many cars. "We wanted to get away from the city rat race, but we're still in it," grumbles Mr. Quill. Says another long-time camper at the same campsite: "This place is skid row now."

The Park Service in Washington is acutely aware of the crowding at Yosemite, and the limit for camping visits may be cut from 14 days to three. Park planners insist there is enough total space in the parks, but complain that at Yosemite and other locations, the public insists on congregating at a few popular, accessible spots. The effects are plain here in the valley.

At dawn and dusk even fresh air is a rare commodity in these crowded campgrounds. Then, when the air is still, smoke from thousands of cooking fires hangs over these areas; ranger Huson takes aspirin to combat frequent headaches when he patrols the "tent tenements," and the heavy wood smoke brings tears to visitors' eyes.

The crowding has had more serious effects. The crime rate has grown so fast at Yosemite that its small corps of rangers now must spend much of their time as policemen. Most are expert foresters with college degrees, and they resent being forced into a law officer's role. Says one young ranger: "A few years ago the ranger was more or less a curator. Today, he's almost exclusively a law enforcement officer. Frankly, I didn't join the park rangers to be a lousy cop."

Ranger Coyt Hackett, responsible for law enforcement in Yosemite, says Yosemite Valley has about the same crime rate as a city of equivalent population (40,000 to 50,000) and adds that it has been rising as the number of visitors rises. The crimes include theft, aggravated assault (leading to an average of about one murder a year), a few sex offenses and a lot of shoplifting. Park authorities collar from 10 to 15 light-fingered visitors in the park's numerous stores each summer day.

Juvenile delinquency has become a major problem. The rangers have secretly formed a "riot squad" to deal with large groups of unruly teen-agers. This group of 20 rangers drills in military formation carrying nightsticks. They were used on one Fourth of July weekend when a crowd of about 300 juveniles blocked traffic on a bridge, shouting insults at passersby and threatening a full-scale riot. The mob dispersed when the special squad arrived.

About a third of the juvenile offenses, however, involve illegal possession of alcohol. During one tour of the campgrounds, ranger Huson investigated a particularly noisy fraternity party, found several underage drinkers, and confiscated a dozen six-packs of beer. This is a familiar chore to the rangers; by Labor Day an office in park headquarters is stacked to the ceiling with the summer's collection of beer and liquor taken from juveniles.

When they are not coping with crime the rangers must devote much of their time to handling a traffic flow that, at times, would give a Times Square traffic cop heartburn. The park has its own rush hour jams; many occur at 9:30 p.m. several nights each week, at "firefall time." The firefall is a huge bonfire that is set high above the valley and pushed off to fall 3,200 feet in a fiery cascade to the valley floor.

There are hundreds of auto accidents in the park each year. Usually, at least one collision a year results in fatalities. The valley floor, ablaze with headlights at dusk, reminds a viewer perched high on its rim of Manhattan's bridges at rush hour.

An art form older than Kilroy adds to the rangers' woes, only today it has a modern twist: Spray cans of black paint have replaced the pocket knife as the traditional instrument used to put initials on rocks, trees, and building walls. Rangers admit they are able to catch few of these culprits.

The crowding at Yosemite also has led to more injuries and accidents to visitors. A few involve hair-raising rescues of climbers, but most are simply "carry-outs"—hikers who have twisted a leg or broken a bone while walking along Yosemite's relatively tame trails. Rangers have worked out a profile of the most accident-prone hiker: A middle-aged, slightly overweight woman who attempts a trail wearing half-heels or loose-fitting shoes. Doctors say this accurately pictures many of the 450 to 500 patients admitted to the 16-bed park hospital each year with broken bones.

The myriad responsibilities of running the park fall on a staff of 97 rangers, only five more than in 1960 despite a one-third increase in park use by visitors since then. Nationally, the park ranger force has increased 55% since 1960, but most of the additional men are filling new posts in recreation areas acquired since then rather than bolstering ranks at older parks, say Yosemite officials.

The rangers at Yosemite have been forced to eliminate some services. Some who formerly patrolled in outlying districts of the Rhode Island-sized park have been pulled in to reinforce the valley staff; ranger stations at the various park entrances are now left unmanned several hours a day, freeing men for valley duties.

"We have nowhere near enough men to protect the visitors and to protect the park," says David Condon, assistant park superintendent. "In the summer, people just take over everything and we can't stop them." Douglass Hubbard, chief naturalist, adds: "The valley is too fragile for the treatment it's been getting. Within a few years we'll have to outlaw both camping and automobiles here—but the guy who first proposes this will have to have guts."

John Hansen, a former ranger and now the park's assistant postmaster, says: "The whole damn valley will be black-topped soon. People today come here to be entertained—they expect popcorn stands and Ferris wheels."

Ferris wheels aren't planned for Yosemite, but the park does have nine grocery and general merchandise stores, seven gas stations, three swimming pools, a coin-operated laundry, a barber shop, a huge stable with more than 300 horses, mules, donkeys and ponies, several bars, and accommodations for 4,500 guests ranging from a room at the luxurious Ahwahnee Hotel (at $49 a night for two, with meals) to a "tent cabin" ($5.50 a day for two).

Most facilities are heavily used. Accommodations at the Ahwahnee Hotel and at Yosemite Lodge are booked solid through the summer. The Ahwahnee, also popular with skiers, is heavily booked during Christmas week; in fact, say hotel officials, you had better make a table reservation now if you want to eat Christmas dinner there 10 years from now.

The park's "village store," a combination supermarket and clothing outlet, grosses $1.8 million annually, and, according to a store official, buys more marshmallows from National Biscuit Co. than any of the firm's other customers—about 2,400 dozen 14-oz. packages a year or about 1.4 million marshmallows. Other volume items: 2,000 can openers a year and 2,500 toothbrushes for the forgetful.

Manning the park concessions in the peak summer months requires a labor force of about 1,500. Many of these are selected by the Yosemite Park & Curry Co., which runs the concessions under a contract with the

ntipollution laws were strengthened by the
gislature, and a new Pure Waters Division was
in the Health Department with orders to crack
rd on all polluters—public and private.

state's program was buttressed by Federal pro-
hat set water quality guidelines for the nation's
ys and promised Federal money to subsidize the
nunicipal sewage treatment.

now it's apparent that there's no chance to
governor's 1972 deadline. State officials say it
975 or 1976 before all industrial and municipal
atment facilities are built—and they may well
uilt even by then.

reason: The $1 billion bond issue simply won't
nough money to finance the state's share of
l sewage plant costs. An unexpectedly high
flation has kicked the price up sharply, for one
d at the same time Congress has failed to au-
nough money to pay for the promised Federal
ion to the cost. Instead of a minimum of 40%,
ayouts in New York have been running about
the state has been obliged to pay the differ-

strial cleanup has lagged somewhat, too.
he worst offenders, state officials say, are a
of paper companies that say they can't—or
end the money needed to process their wastes.
some—including Finch-Pruyn and Patrician
re going ahead with treatment facilities, oth-
hreatened to close down operations instead.

conservationists and scientists who study the
they are less worried about the delays in the
than about the possibility that the technical
e needed to clean up the river simply doesn't
y point, for instance, to the inability of even

Park Service, from the ranks of college students applying for summer jobs. A far smaller force—about 200 blue-collar Park Service employes—keeps the rest of Yosemite humming.

The huge park is almost entirely self-sufficient. The workers operate a hydroelectric power generating station that delivers 16 million kilowatts a year, a sewage system that treats 268 million gallons a year and an incinerator that burns 2,100 tons of garbage a year. Trail crews, some of which remain in the wild back country for weeks on end, maintain 354 miles of roads and 768 miles of trails. Five men driving tiny motorized trucks called "litterbuggies" manage to keep Yosemite Valley remarkably free of debris; due to limited manpower, however, they can spend little time outside the valley—so beer cans and pop bottles sometimes lay uncollected in the scenic upland areas.

The park's operating budget has risen steadily, but park officials say that budget increases in recent years have been barely enough to cover salary boosts, and they fear that a continuation of the tourist explosion will place even greater strain on the park's ability to handle its visitors.

Only Yosemite's bears seem unworried by the crowding. Indeed, they've become so spoiled by copious handouts they frequently turn on tourists who offer what the bears evidently consider niggling amounts of food. Most injuries are caused by visitors who either pretend to offer something to eat, or throw the bears a sandwich. "That's like giving a child one potato chip," says ranger Hackett. In any case, the park doctors treated about 50 people who were injured by bears last year.

—GLYNN MAPES

1966

The Hudson: A Majestic Mess

THE view of the Hudson River from the bridge at Glens Falls, N.Y., may not rank among the world's scenic wonders, but it certainly is colorful.

On the left as you face downstream, a bright turquoise-colored fluid spews into the river from a waste outlet at the big Finch, Pruyn & Co. paper mill that stretches for hundreds of yards along the river. At other times, the waste may be almost any other color, depending on what hues the company is dyeing its products on a given day.

To the right, tumbling down a bluff that rises along the opposite bank of the river, is a spillway that carries waste from the Patrician Paper Co. plant high atop the bluff. The spillway channel is deeply stained from the dyes that have coursed through it to the river over the years. It is red in some places, dark blue-green in others.

Paper makers use colors, too, to describe other, less obvious types of effluent that many plants dump into the river. There is, for instance, the "black liquor" that comes from pulp mills—a dark, smelly mixture of acids and residue from the breakdown of wood pulp into paper fibers. And then there's "white water," a discharge containing tiny fibers that escape the machines

that press fibers into paper
the river, the waste fibers
eventually rot into a putrid

In all, there are some
son, and they are far from
handy means to get rid of
State Health Department
polluters just in the lower
Troy and New York City—
other institutions ranging
U.S. Military Academy at
dump close to two billion g
river every day. Most of it
carries with it a staggeri
acids, oil, silt, soap, deterg

The pollution of the
that it has made the river
lution fighters—and it ha
bitious river cleanup effor
try. But it has also rais
among conservationists,
who wonder if even the m
paigns can restore the p
has been so drastically al

In 1965, New York's
that if the state's voters
waters" bond issue to h
plants throughout the
would get first priority.
be done.

The bond issue pas
margin, and the state p
of seeing that all of Nev
pal wastes undergo trea
into the Hudson and th

state.
state le
created
down h

Th
grams
waterw
cost of

Bu
meet th
will be
waste tr
not be

On
provide
municip
rate of i
thing, a
thorize
contribu
Federal
7%—an
ence.

Ind
Among
number
won't—s
Though
Paper—
ers have

Still
river say
timetabl
knowled
exist. Th

"modern" sewage treatment techniques to remove such nutrient materials as phosphorus and nitrogen from waste water. The nutrients could eventually cause the river to become so choked with algae that it might be far less attractive and far less able to support marine life than it is now.

A report by scientists studying the ecology of the Hudson for the Institute of Environmental Medicine, part of the New York University Medical Center, warns that a number of "nuisance" species of algae already exist in the Hudson. Moreover, the report says, nitrate and phosphate are already present in quantities great enough to foster disastrous algae growth under certain conditions.

"The present move to treat sewage so as to provide a liquid effluent . . . could, in fact, worsen the present situation since treated effluents provide nutrients in a more readily available form," the scientists warn.

When explosive algae growth occurs, the water turns green and develops a scum on the surface. As algae die and rot, they use up oxygen. Eventually, the water is so robbed of dissolved oxygen that it's unable to support normal marine life. The process is known as eutrophication.

Some scientists and conservationists are particularly concerned that eutrophication could be set off in the Haverstraw Bay area about 30 miles north of New York City by a cluster of nuclear power plants being built just north of there. Such plants use millions of gallons of water a minute for cooling and return it to the river some 15 degrees hotter than the normal water temperature. There's a danger, conservationists contend, that the slight resulting rise in overall water temperature could trigger an algae population explosion; algae seem to thrive in warm water.

"All of us are aware that more power plants are going to be built, and most of us don't want to go back to darkness," says Rod Vandivert, executive director of a conservation group known as the Scenic Hudson Preservation Conference. "But we're aware that there's a very delicate balance that's in danger of being upset, and there's a hell of a lot of us that don't feel government is doing its job in allowing these plants to be built so close together."

Consolidated Edison Co., which has one nuclear plant unit in operation in the Haverstraw Bay area and plans to build four more within a one-mile stretch, says its discharges comply with state criteria for heat. Moreover, a spokesman says the company's studies indicate that, in the Haverstraw Bay area at least, the resulting water temperature rise isn't likely to cause eutrophication.

This difference of opinion points up another problem: A lack of research on aquatic ecology. "The thing that's important and frightening is that we're working on a series of best engineering guesses with no real marine research behind them," says Mr. Vandivert.

Removal of nutrients might be accomplished by so-called tertiary treatment of waste—a step beyond the customary primary and secondary methods of treatment that are included in the sewage plants now being built. But New York State Health Department scientists say no economical and efficient method of tertiary treatment has yet been devised, though the department's research staff is working on what is described as a promising method of chemical treatment of waste water.

Another possibility is the limiting of nutrients by removing them from laundry detergents, which Health Department officials say are the chief source of nitrate and phosphate in the Hudson and many other waterways. Rep. Henry S. Reuss, a Wisconsin Democrat, has introduced a bill in Congress that would ban phosphate from detergents within two years, but its chances of passage seem dim for the present.

Conservationists' concern for the Hudson extends well beyond the question of water quality. They argue that the Hudson Valley possesses unique scenic and historic qualities that are constantly endangered by various development schemes, and they complain that Government agencies have failed to move forcefully to clean up the river's often unsightly shores and to protect the natural wonders that remain.

Several times in recent years, enraged groups have stepped in to battle what they regarded as threats to the Hudson or its valley. The highly publicized fight to prevent Consolidated Edison from building a pumped-storage hydroelectric plant at Storm King Mountain, for instance, is being waged chiefly by the Scenic Hudson Preservation Conference, an amalgam of individuals and conservation groups that has spent more that $500,000 in the past five years fighting the plant. (At the moment, the battle is undecided; Con Ed has agreed to drastically alter its plans, putting the plant underground, but Scenic Hudson is trying to persuade the Federal Power Commission to block it altogether as a general threat to the area's natural environment.)

A particularly difficult problem for those who would improve conditions along the Hudson is the river's relative isolation from the people who live near its banks. Besides the railroad tracks that run up both sides of the river much of the way from New York to Albany, factories and "farms" of oil storage tanks crowd the shores in most built-up areas. Inland from the factories and railroad tracks in many old Hudson towns, moreover, are bands of slums or decaying old commercial buildings.

Altogether, of course, these conditions tend to discourage residents from seeking out the water for recreation—or from taking much interest in whether it's polluted or not. Folk singer Pete Seeger has tried to stir up more interest through a group known as Hudson River Sloop Restoration Inc., which built and sails a sloop of the type that once was common in the Hudson River trade. Mr. Seeger and other entertainers ride their sloop, the Clearwater, up and down the river during the summer months, stopping to give concerts and exhibit the boat as a means of attracting publicity and getting valley residents concerned about the river.

In 1965, Gov. Rockefeller created the Hudson River Valley Commission to draw up a proposed master plan for the valley and to review all projects along the river with an eye toward protecting the scenic, historic and cultural resources in the area. Conservationists complain that the commission lacks enforcement powers; it can only "recommend" changes in a project. But its persuasive powers have, nonetheless, managed to forestall the building of a nuclear power plant directly across the river from the Saratoga National Historic Park, get another power plant's design changed to meet aesthetic objections and save a row of historic houses that were to be torn down in Albany.

The scenic qualities of the Hudson have inspired travelers on the river for centuries. Early Dutch explorers 300 years ago rhapsodized about the great forests and mountains lining the river, the sweetness of the air, the abundance of game and the fish that teemed in the river. In the 19th Century, the Hudson was a fashionable tourist spot—as well as the inspiration for the Hudson River school of painters, the first American artists to devote themselves to landscapes rather than portraits.

Such wonders as the Hudson offers are compressed into a length of merely 315 miles, making it one of the shortest of the nation's major rivers. It starts at Lake Tear of the Clouds, a pond on Mount Marcy in the Adirondack Mountains of upstate New York, and flows southward to empty into the Atlantic at New York Harbor.

The upper Hudson is a fast-flowing trout stream that plunges through a deep, forested gorge in the Adirondacks. Not until 150 miles downstream, after it is joined by the Mohawk River just above Troy, does the Hudson begin to resemble the broad, majestic waterway that Henry Hudson named the Great River of the Mountains when he explored its lower half in 1609.

After flowing past the Catskill Mountains to the west, the river enters the Hudson Highlands, narrowing into a gorge flanked by steep mountains with peaks called Storm King, Breakneck Mountain, Bear Mountain, Anthony's Nose and High Tor.

Then abruptly, it widens to its broadest areas—Haverstraw Bay, some three and one-half miles across, and the Tappan Zee, almost as wide. A few miles above New York City, the river narrows again and flows past the Palisades—sheer rock cliffs that rise dramatically almost 500 feet above the west bank.

The lower 150 miles of the Hudson actually constitute a largely saltwater estuary of the Atlantic, where tides ebb and flow. Because it is at sea level, with no falls or rapids, and because it's so wide and deep (more than 100 feet in some areas), this stretch of the river is navigable to ocean-going ships all the way to Albany.

Significant pollution of the Hudson begins at Corinth, N.Y., some 60 miles above Albany. There, International Paper Co. operates the northernmost of the paper plants on the river. The effect of the plant's waste discharges is evident for miles downstream.

It's possible, for instance, to reach into shallow water along the shore below Corinth and tear chunks of thick, gray, paper-like material from the river bottom. The material is part of a dense mat created by waste fibers that have settled to the bottom.

On hot summer days, when heat speeds up decomposition of the fibers, the placid surface of the river below Corinth is periodically disrupted by the bursting of giant gas bubbles. The bursting of a bubble is typically followed by the appearance of a great chunk of matted fibers that has been loosened by the decomposition and has risen to the surface, where it bobs off like a tattered raft.

In such areas—which extend for many miles along the river in the paper-making region around Glens Falls—the river cannot support normal aquatic life, scientists say. The matted fibers covering the bottom choke off plant life, and the decomposition of the fibers and other paper mill waste robs the water of dissolved oxygen needed by fish. The only organisms that thrive in those conditions are such creatures as sludge worms and maggots that feed on the decomposing material and seemingly require almost no oxygen.

Once past the paper mills, the river's natural ability to cleanse itself has barely had a chance to function before it hits the Troy-Albany metropolitan complex, where a half-dozen cities, many of them heavily industrialized, cluster along both banks for a 10-mile stretch. This part of the river is known to river experts as the Albany pool—but sanitary engineers wryly refer to it as the "Albany cesspool."

The reason for the nickname is obvious. Scores of

cities and industrial concerns dump raw or partially treated sewage into the river within the few miles on either side of Albany. Factories dump acids, drugs, dyes and chemicals. Oil barges, back in the Hudson after delivering oil up the Mohawk, furtively discharge ballast water they have taken on for the return trip down the Mohawk. Because the ballast water has been in the cargo tanks, it carries a residue of oil.

A walk along the river bank in the Troy-Albany area gives some indication of the result. The water is dark and murky-looking, the surface frequently rainbow-hued with oil slicks. Bits of trash bob along the surface, and quiet backwater coves along the shore are jammed with debris—rotting boards, empty steel drums, tires and old shoes. Rocks on the banks are coated with a black, sticky gunk, chiefly oil residue.

Perhaps the most notorious sewer outflows on the Hudson are at Troy. Like most river towns, Troy is built on hills that slope down to the water. Its sewers simply run under the streets leading to the Hudson. At the base of each street is an outfall—a pipe jutting through the concrete retaining wall along the river.

At low tide, the outfalls are exposed as they dump raw sewage directly into the Hudson. Even at high tide, you can stand on the retaining wall and watch the gray, filth-laden sewage water bubble to the surface.

Between Albany and New York City—a distance of some 150 miles—pollution is relatively light. Scattered towns and factories dump wastes directly into the river, but the Hudson's volume is so great that it's able to absorb them and still cleanse itself to a large extent. Public health authorities permit swimming at Senasqua Park in Croton-on-Hudson, a northern suburb of New York, and fishermen say there's no shortage of fish in the river—though some complain that they taste a bit oily.

Robert H. Boyle, an avid fisherman and conservationist who wrote a recent book on the Hudson, says: "As of now, the biological productivity of the lower Hudson is staggering."

But in the New York City area, the river once again becomes an open sewer. Manhattan still dumps practically all its sewage directly into the Hudson and the other waterways that surround the island, and a number of communities across the river in New Jersey also pour waste into the Hudson with little or no treatment. Researchers at the New York State Health Department calculate that 1.6 billion gallons of municipal and industrial waste is dumped into the Hudson daily in the New York area. More than 90% of the industrial waste and close to 70% of the municipal waste has received either no treatment or only minimal treatment.

Because of the tidal nature of the lower Hudson, sewage from New York may actually wash upstream as far as 50 or 60 miles. Then it may slosh back and forth with the tides for weeks before it finally escapes into the Atlantic.

Altogether, the Hudson's pollution load seems gigantic, but state Health Department officials contend that they will be able within the next few years to eliminate the most obvious types of filth. Though the state is running short of money to help municipalities build sewage treatment plants, the priority assigned the Hudson has assured that most sizable cities on the river will get state help nonetheless.

Similarly, officials say they are determined to force industrial concerns to clean up their waste. Dwight F. Metzler, the Health Department's deputy commissioner for environmental health services, says: "Industry is told, 'We don't care how you get rid of this. You can have your employes drink it, you can put it into a tank car and ship it to Canada, you can incinerate it—that's up to you. All we care about is what comes out the discharge pipe.' "

As a practical matter, however, solutions aren't likely to be forthcoming quickly. Paper companies say, for instance, that much of the technology for treatment of their wastes is so new that extensive engineering work is necessary before they can determine what equipment is needed.

State Health Department officials contend that at least some of the delays are foot-dragging. Andrew M. Weist, a Health Department water pollution control engineer whose region includes most of the Hudson's paper plants, says some paper companies have managed to postpone commitments for years by pleading that they need more time for study.

"They want six months for this, a year for that—and still nothing gets done or even firmly committed," says Mr. Weist. "You sit there and you just keep pounding your head against the wall."

Not all companies have resisted, to be sure. Finch-Pruyn has already installed some $6 million worth of new equipment that will make a "drastic reduction" in the black liquor discharge from its pulp mill, a spokesman says. The company, meantime, is drawing up plans for cleaning up its other effluent.

Patrician Paper has designed a new high-efficiency filtration system that it says will cut its effluent by 80%.

International Paper says it has begun work on detailed construction plans for its treatment facilities, estimated to cost $2.8 million, but a company spokesman says construction won't be completed for close to two years.

Westvaco Corp., which operates a pulp and paper mill at Mechanicville, between Albany and Glens Falls, complains that full treatment facilities for its plant would cost an estimated $3.5 million, more than the book value of the 80-year-old plant. They add, however, that they expect to meet the current deadline of March 1, 1971, for submission of a final plan.

—JOHN BARNETT

1969

The Mustangs

THE wild bay stallion panics as the strangers approach. Eyes rolling, he races desperately around the corral that is his prison, tripping over wires, crashing against the fence.

He had come down from the arid mountains near Caliente, Nev., and was drawn to a familiar salt lick that had been fenced, its gate left open. Entering, he tripped a wire and the gate slammed behind him. When the gate opened again it would lead to the back of a truck. There would be other corrals then, and, finally, the last stop at a pet-food plant.

Trapped and sold for six cents a pound on the hoof, shot and left to rot on the range, driven off cliffs, the mustang is a slowly vanishing symbol of the American West. Descended from stock brought to the New World by the Spanish conquistadors and the settlers who came after, mustangs once roamed the Great Plains by the hundreds of thousands. No more.

As barbed wire and the plow marched across the grasslands, the mustang retreated farther and farther west. Exterminated as a pest by ranchers, driven into dry country where pasturage is hard to find, his numbers dwindled. Today about 17,000 mustangs are left in 11 Western states, about half of them in Nevada, many in the arid moonscape of that state's southern region.

Conservationists—and some repentant "mustang-ers" who have slain many of the animals in the past—are now working to preserve those that are left. They will have to work fast; Stan Routson, Nevada state inspector of brands, expects the mustang to survive only about eight years more in his state.

"The mustangs are definitely going down the drain," says Tom Holland, a retired Utah rancher who heads the National Mustang Association, which wants to set up a 68,000-acre preserve near here. "Horses helped build and conquer America, and the wild ones left on the range ought to be a national monument. Instead they're treated like garbage, like so many empty beer cans."

If the mustangs do disappear, it will be because of economics and the animal's lack of status as an endangered species. The Federal Government doesn't recognize it as such because, strictly speaking, the mustang is not a distinct species, just a horse of mixed parentage. Old mustang hunters, however, view him as a wild animal in the truest sense of the word; they have seen mustangs leap over cliffs rather than be caught or starve themselves to death when trapped in corrals rather than eat the food proffered by captors.

Mrs. Velma (Wild Horse Annie) Johnston, a Reno secretary who has been a strong champion of the mustang, says: "No one wants to be responsible for them. They're not an edible animal and they're not a trophy animal. They're in limbo."

Mrs. Johnston was instrumental in obtaining a 1959 Federal law that forbids the use of planes and motor vehicles in rounding up mustangs on Federal land (86% of Nevada is Federally owned). A bill introduced in Congress earlier this year would give the Secretary of the Interior authority to set up and maintain preserves for mustangs and burros, but the fate of that is uncertain and the 1959 law is the principal protection now available.

The 1959 law has ended the massive roundups of the past, when hundreds of mustangs were driven over cliffs or rounded up and sold by teams of hunters in jeeps, trucks, and airplanes; about 100,000 of the wild horses were removed from Nevada in the four years after the end of World War II, for example. Still, the 1959 law is poorly enforced and has holes in it.

"You're not going to get the FBI out here to enforce it," says Mr. Routson, the Nevada brand inspector. "And the county commissioners are elected by ranchers, who want that land for cattle."

Many ranchers lease grazing rights from the U.S. Bureau of Land Management, paying a fee based on how many cattle the bureau estimates can be supported on a given tract. In the Nevada badlands, where cattlemen joke that a cow has to graze at 20 m.p.h. to get enough to stay alive, every tuft of green is precious—and the mustang is a wide-ranging rival for available pasturage.

So ranchers and others still trap and kill mustangs legally and illegally. The anti-motor vehicle provision of the 1959 law is evaded by mustang hunters who release their own branded horses into a mustang herd—and then chase and round up the whole lot using jeeps and trucks. If anyone gets curious, the hunter can argue that he was just after old dobbin.

There is some evidence, though, that as the mustang population has dwindled, the ranchers have softened. Says Walter Clutts, head of the Las Vegas chapter of the National Mustang Association: "Fellows who rode them off cliffs by the hundreds are starting to support our program. They look out the window and they don't see them anymore. Deep in their hearts they love them and they want to see some around; it's hard for a rancher not to have some feeling for horses."

Ellis (Cougar) Lefevre, a grizzled mustanger who says he was personally involved in the capture and extermination of some 10,000 of the animals, says he repented years ago when it struck him that his children might never see a mustang. Retired rancher Holland, the Mustang Association president, recalls a similar change of heart.

"Back in the fifties," he says, "there was one horse, a white mare. We chased her for three years. We got to calling her Native Dancer. The last year we chased her, she was with a colt. We finally caught her, but it took three relays of men and horses and an eight-mile run. Now, that horse was the goingest concern I ever come across, and it about broke my heart when we shipped her out for pet food."

—P. F. KLUGE

1970

Land for Posterity

THE Potomac River ice creaks and groans beneath the January sky. Cardinals flit across the beige and white of the snowy cattail marsh and crows caw from nearby woods of beech and oak. A great blue heron lifts away on three-foot wings.

Mason Neck on a clear, cold morning is placid, unhurried now. But in 1965 this 10,000-acre peninsula on the Potomac in Virginia was threatened by the relentless spread of suburban Washington. Real estate speculators controlled the land; there were plans for asphalt streets through the woods, subdivisions near the restored mansion of a Colonial planter.

It didn't happen. And the main reason was the quiet work of an increasingly effective conservationist group, the Nature Conservancy.

In 1967, the Conservancy moved in and began buying up more than 3,000 acres for about $5.6 million, checkerboarding its holdings to block development of most of the peninsula. It was another successful application of one technique that helps make the Conservancy unique among national conservation groups—unique in what it does as well as what it doesn't do.

The Conservancy isn't the best-known national conservation organization. It rarely makes headlines with dramatic protests or last-ditch lawsuits. It doesn't sponsor wilderness outings and it doesn't publish beautiful books.

It just preserves land, the kind of land that can't be replaced: Virgin woods in New Jersey, islands off the Atlantic Coast, ancient California redwoods, prairies, marshes and mountains. The Conservancy is the only national conservation group that puts its total resources into land preservation. So far, it has preserved about 150,000 acres in 41 states and the Virgin Islands—most of this since it really got rolling in the early 1960s.

The Conservancy traces its lineage to a 1917 committee formed to acquire natural areas for scientific research. Today, however, the Conservancy is interested in outstanding examples of the American environment for other purposes as well. It buys such land itself or lends money to private groups that wish to do so; tax-exempt and nonprofit, it accepts bequests and donations of land or cash. It has helped preserve everything from a 10,500-acre island off Georgia (now a Federal wildlife refuge) to Ezell's Cave, the subterranean home of *Typhlomolge Rathbuni,* the Texas blind salamander.

Both public and private efforts to preserve natural areas threatened by development often founder for the same reason: A lack of ready cash. By the time a government agency can secure its appropriation or a citizens group can launch a fund-raising drive, the bulldozers have come and gone. The Conservancy is trying to fill this gap with three programs:

—From a revolving fund of more than $1.1 million, it makes quick loans to private groups, including its own chapters, organized for the purpose of acquiring specific areas. The groups may take up to three years to repay; the loans are interest-free for three months, then bear interest at an annual rate of 6½%.

—A separate endowment fund of about $800,000

guarantees bank loans to such groups when the revolving fund is being used to capacity.

—Under its newest program, which utilizes a $6 million line of credit guaranteed by the Ford Foundation, the Conservancy moves in fast to acquire tracts being sought (for parks or wildlife refuges, for example) by Federal, state or local government agencies. It resells the land to the agencies when their appropriations come through.

Requests for help are keeping all three funds busy. A loan to a citizens group, for example, recently helped preserve Clausland Mountain, a wooded rampart on the Hudson River near New York City. The $237,500 loan clinched offers of more than $1.1 million in additional money from other sources. Area artists have raised some of the money for repayment with an "Art for the Mountain" benefit.

The program using the Ford-guaranteed credit line has acquired more than 11,000 acres since early 1969, sometimes nailing down tracts that slower-moving government agencies might have lost. A good example is the 3,215 acres of Michigan forest recently acquired for the U.S. Forest Service. The Federal agency turned to the Conservancy because the tract was being marketed by a concern that needed to sell quickly, and it might have taken the Forest Service as long as 18 months to secure the necessary appropriation.

Such successes are winning the Conservancy support from figures as diverse as Laurance Rockefeller, Charles A. Lindbergh, Arthur Godfrey ("Boy, they do a job") and Marshall Field. Says a top Federal conservationist: "They haven't tried to branch out and get involved in all aspects of the environment. They've stuck to land preservation—and they're doing it damned well."

Conservancy officials praise the efforts of such better-known organizations as the Sierra Club, which attempts to rouse public opinion and sometimes hauls developers and polluters into court. But the Conservancy generally avoids such fights. "The measure of our success is not how well we propagandize for or against a given issue," says Thomas W. Richards, president. "It's in those acres, and in the quality of those acres."

So it's no accident that Conservancy headquarters in downtown Washington rather resembles a high-powered real estate agency. It's the kind of place where Mr. Richards may interrupt an enthusiastic description of a contemplated project (enclosing both banks of a portion of the Potomac in a "green sheath," for example), to answer the telephone and bargain for an island, a marsh or a forest. The atmosphere seems a little like that in a cartoon above the desk of Edward R. Kingman, vice president and treasurer. The cartoon depicts an exasperated executive who bellows: "Whattya mean we don't have any capital. . . . The acquisition's already been approved."

The cartoon notwithstanding, the Conservancy is at home in the world of finance. Mr. Kingman has been a bank vice president, a financial consultant and a real estate broker; Mr. Richards has nine years of experience as an IBM department manager. Other staff members include ex-real estate agents, a NASA administrative assistant and an industrial engineer—all recruited for their management skills.

"Conservation problems today are no longer solved by a guy hiking around in the woods," says Alexander B. Adams, an ex-FBI agent who helped lead the Conservancy through most of the 1960s. "They're solved by guys sitting behind desks, thinking." Agrees Mr. Richards: "To win a land conservation battle today, you've got to use the same skills private industry uses."

In 1969, its biggest year yet, the Conservancy helped preserve nearly 40,000 acres through 101 projects and donations. The year also marked ceremonial completion of a major phase of the Conservancy's most spectacular project to date: The addition of about 10,000 acres to Hawaii's Haleakala National Park.

Before the project, Haleakala Park occupied about 14,000 acres atop a long-extinct volcano. Soon the park will contain about 24,000 acres and extend from the mountaintop to the sea, an enlargement that one conservationist calls a "dream come true." It all began with a 1967 challenge from Laurance Rockefeller. He would donate a $585,000 piece of shorefront to the park—if the Conservancy could acquire the eight-mile-long Kipahulu Valley between the shore and the mountaintop.

Often veiled in fog or drenched in torrential rainfall, the valley is a lush remnant of Hawaii as it used to be. More than 100 waterfalls roar in a rain forest abundant with wildlife, including a bird species presumed extinct for 80 years. The upper valley is a wilderness scarcely penetrated by modern man. Not surprisingly, the Conservancy took the challenge and went to work.

As negotiator, the Conservancy dispatched Huey Johnson, its western regional director. In two weeks of hectic bargaining, Mr. Johnson reached agreements with the valley's three private landowners, then persuaded the state of Hawaii to donate about 3,000 additional acres it held.

The private owners eventually sold nearly 7,000 acres for $620,000, donating additional acreage valued at $300,000 as a tax-deductible contribution. A mail solicitation, three cocktail parties and a luncheon raised the $620,000, with about $375,000 coming from a gathering in New York's Pan Am building. Mr. Lindbergh addressed that gathering, and Mr. Godfrey did a persuasive job, too. He describes catching a departing donor in the elevator and emerging at the end of the ride with a pledge of $100,000.

In January 1969, the Conservancy donated more than 7,000 acres to the National Park Service under an agreement that will preserve the upper valley as wilderness for scientific research and open the remainder of the valley to the public. (The state is in the process of conveying its 3,000 acres to the Park Service.) Then the Conservancy launched the project's second phase: A campaign to raise about $750,000 to purchase several hundred additional shorefront acres highly vulnerable to development. If this phase succeeds, Gov. John Burns has indicated, he'll work for the donation of additional state land. Says Mr. Richards: "We want to do this thing once and for all, and do it right."

The scope and expertise of the Kipahulu project was a far cry from the Conservancy of 1960. That year the organization preserved only about 4,000 acres, had an operating deficit and only about $100,000 in its revolving loan fund, and was mired in an ill-planned project that threatened to bankrupt it. Adds, Mr. Adams, then president: "We were like practically every other conservation group—trying to do everything at once, and not doing anything as well as we might."

Spurred by Mr. Adams, the Conservancy reorganized. It beefed up its staff with the help of Ford Foundation grants, formed the endowment fund and secured the Ford-guaranteed line of credit. And after what Mr. Adams calls "a long battle within the organization," it phased out activities unrelated to land acquisition.

This meant leaving public protests to other conservation groups, a decision that still has its critics. One, for example, asserts that "too much concern about what major contributors might think" sometimes inhibits Conservancy activities and was a major factor in the policy change.

This critic is particularly disturbed because in the early 1960s the Conservancy dropped an active role in opposing a controversial pumped storage hydroelectric plant proposed by Consolidated Edison for New York's Storm King Mountain. He maintains: "Many Conservancy backers are stockholders of Con Ed or are interested in other forms of economic development along the Hudson and might have been offended."

Mr. Adams disagrees. "I know of no instance where our policy has been affected by a donor, and I can say that absolutely flatly," he declares. He calls the protest against the Storm King plant "the kind of project that could be much better handled by other groups" and notes that another group did take over after the Conservancy dropped out. The intent, he says, was to "disengage from things other organizations were already doing and concentrate on buying land."

There's no doubt that Conservancy fortunes soared after the reorganization. In 1969, it either bought or received as gifts land valued at nearly $20 million, up from about $750,000 in 1960; by 1975, it expects this amount to rise to $50 million. During 1969 it transferred ownership of $7.2 million worth of land to various Federal, state and local institutions, including universities.

Increasingly, the Conservancy is going into large-scale projects that will protect complex life chains in broad areas. A top priority for the 1970s will be the acquisition of coastal marshes and wetlands to protect spawning grounds for marine life and refuges for migratory birds. Separate projects, already well under way, aim to establish "coastal reserves" of islands off Georgia, Virginia, Maine and Florida. Other priorities: The acquisition of virgin prairie, water-filled "potholes" (needed by migrating ducks and geese) in the upper Midwest, and desert springs and streams.

In 1970 the Conservancy planned to spend $7.5 to $10 million for land acquisition—a record but about $31 million short of what it would like to spend, says Mr. Richards. He estimates he would need at least $15 million more, for example, to buy up "some of the most critical inholdings" (private land) within national parks and other public areas; $10 million more to fully execute a new project to protect threatened wetlands around San Francisco Bay; $3 million more for Gulf Coast Florida islands and wetlands; and $3.5 million for Atlantic barrier islands and salt marshes.

Meanwhile, additional requests keep coming in. Illinois is asking help in buying a $7.8 million piece of open space in Chicago, for example. And Sen. Ralph Yarborough (D., Tex.) has asked for help in preserving something of East Texas' Big Thicket, a beautiful forest of pines and hardwoods.

Private donations and fund-raising drives by Conservancy chapters and project committees brought in nearly $5.5 million in cash and securities in 1969. Donors also contributed about $12.5 million worth of land, including a 74-acre ridge in Connecticut and 361 acres of forest (valued at $1 million) in Florida.

"We're willing to go to almost any lengths for a donor," says John F. Jaeger, the staff attorney who processes most of the gifts and bequests of land. Some donors retain the right to live on the donated property for their lifetimes, for example. Others donate only a portion of the value of their land and sell the remainder to the Conservancy, or assign ownership to the Conservancy over a 20-year period.

The Conservancy is looking for help from another area: Business. In 1969, in what Mr. Richards called a "breakthrough for conservation," the Conservancy accepted a gift of two groves of California redwoods (worth about $6 million) from Georgia Pacific Corp., a concern that drew bitter attacks from some other conservation groups during the fight to establish the new Redwoods National Park. The gift, now a California state park, convinces Mr. Richards that business and the Conservancy can work together wth mutual benefits.

"I'm anxious to work with other businesses, particularly the extractive industries," he says. "It's conceivable, for example, that a lumber company could assess its massive holdings and find some areas that aren't beneficial to it but which would be great from our standpoint. We could take management problems off their hands and enhance their public image in the process."

It's an irony of Mr. Richards' work that he seldom escapes his office to visit the landscapes he's helped preserve. (His most satisfying acquisition to date is a Georgia island he has yet to visit.) But he's an enthusiastic outdoorsman, as a winter hike on Mason Neck well indicates.

A jaunty beret on his head and field glasses swinging from his neck, Mr. Richards strolls across the iced-over marsh and into the woods, checking tracks in the snow and training the glasses on birds that wing by. "Boy, isn't that great!" he exclaims, focusing in on a flying woodpecker—red and white and black against the sky. Still watching, he quips: "Look at that body!"

He studies a distant treeline, the last known nesting area of the bald eagle on this stretch of the Potomac. (The marsh and nesting area, part of the acreage acquired by the Conservancy, will soon be a Federal wildlife sanctuary; other tracts on the peninsula will become state and regional parks.) Then it's on to Gunston Hall, the restored mansion of George Mason, a close friend of Thomas Jefferson. Residential subdivisions had been planned near Gunston Hall before the Conservancy intervened.

Later, in the formal garden behind the red-brick mansion, Mr. Richards stops to savor the view: The 200-year-old hedge of English boxwood, the giant oaks, the uncluttered woods beyond.

"This will give you an idea why the Conservancy is here at Mason Neck," he says. "We're not just saving a bald eagle sanctuary. By God, this is part of this country's heritage, and it shouldn't be messed up."

DENNIS FARNEY

1970

The Army's Embattled Engineers

WHEN the U.S. Army Corps of Engineers began work on a flood control project in Stuart, Fla., some 20 years ago, alarmed townspeople tried to halt it. They feared ruination of their picturesque river, the St. Lucie. But the Corps assured residents its project would enhance boating, swimming and fishing, and the project went ahead.

The people were right and the Corps was wrong.

Today the once-clear St. Lucie is black with mud, and Corps officials in Florida admit their agency is largely to blame. Nearly all the fish are gone. Gone, too, are most of the oysters, clams, pelicans, ospreys and wild ducks. Now Corps officials talk of a costly new project—a 10-year effort to clean up the river.

The venerable Corps is under attack in this Atlantic coastal town, as it is across the nation whose waterways it has often dammed, rechanneled, dredged or filled in. Critics range from ordinary citizens to U.S. Senators and Supreme Court Justice William O. Douglas, who calls the Corps "public enemy number one." They charge the agency with bureaucratic high-handedness and assert that many of its expensive projects are not boons, but boondoggles. But their primary accusation goes deeper than that.

An increasing number of critics contend that many Corps projects wreak environmental havoc, disrupting nature's intricate cycles at a cost that is not only aesthetic but also economic. They say the Corps has an "engineering mentality" that worries only about the mechanics of its projects and couldn't care less about the side effects.

Corps officials bitterly resent such charges. "Those silly butterfly chasers and self-serving politicians can't stand in the way of progress," snaps one staff official. A Corps spokesman buries his head in his hands and mutters softly: "Those ignorant, misguided, conceited fools, they know not what they say. We are the nation's leading conservationist group because we have conserved the earth by molding it to suit man."

The Corps marshals statistics and colorful brochures to justify the approximately $21.5 billion it has spent on civil works projects alone since the initial project in 1824. The Engineers estimate, for example, that the approximately $5.5 billion they've spent on flood control projects since the late 1920s has prevented some $15.5 billion in damage.

Elaborate Corps canal projects have made inland ports of landlocked cities like Tulsa, Okla., unquestionably boosting regional economies. Giant flood-control reservoirs have been a boon for boatsmen and water skiers. And even Corps critics find something to applaud in some Corps projects, like the aid the agency gave recently to the hurricane stricken Mississippi and Louisiana Gulf Coasts.

After Hurricane Camille struck Gulfport, Miss., the Engineers moved in quickly to help rebuild public roads, bridges, buildings and harbors—and the restorations often were better than the originals. "God bless the

Army Engineers," says Philip Shaw, mayor of Gulfport. "They have helped resurrect our stricken city."

The Corps can trace its lineage back to 1775, a history that spans war and peace. In wartime, Engineers build roads and bridges, plant mines and booby traps, clear debris and, if necessary, fight. (The Corps has 35,000 combat engineers in Vietnam.) At the height of its activities during World War II, the Engineers oversaw construction in the U.S. at the rate of $20 million a day and supervised the "Manhattan Project," which developed the atomic bomb.

Domestically, the Engineers are responsible for flood control, improvement of navigable rivers and harbors and national water resource planning and development. Today the Corps spends about $1 billion a year, employs 10,000 officers (including an elite contingent picked from the top 30% of West Point graduating classes) and 50,000 civilians. It also oversees projects that involve another 50,000 civilians working for contractors. It is a potent force, economically and politically.

Because its allies on Capitol Hill are legion, the Corps has ridden out many a past attack. Harold Ickes, Interior Secretary under Franklin D. Roosevelt, once wrote: "It is to be doubted whether any Federal agency in the history of this country has so wantonly wasted money on worthless projects as has the Corps of Engineers." Presidents Roosevelt and Johnson both considered trying to limit the Corps' power, but neither managed to do so.

But the current attack may be different. Corps foes are united and are buoyed by growing national concern about the environment. Conservationists have become increasingly adept at mobilizing public opinion and using lawsuits to force the Corps to leave nature alone.

Some recent conservationist victories:

—A conservationist campaign in 1968—plus intervention by Gov. Louie B. Nunn and Sen. John Sherman Cooper—persuaded the Corps to modify a proposed dam that would have flooded the most scenic part of Kentucky's Red River Gorge.

—Public outcry against a plan by Coastal Petroleum Co. to mine limestone from Florida's Lake Okeechobee induced the Corps to deny a necessary permit. Opponents had feared the plan would ruin the 750-square mile lake for recreation and heavily damage fish and wildlife.

—Lawsuits filed by conservationists invalidated a Corps permit allowing a landfill project along the Hudson River and compelled the Corps to start enforcing a law prohibiting pollution of the river.

—Complaints against the controversial and partially completed $160 million Cross-Florida Barge Canal, plus a lawsuit against the project, have encouraged a state legislative committee to announce a full-scale investigation.

"We're getting swiped at in lawsuits all over the nation," says one Corps spokesman.

Corps projects have also been spotlighted by a growing number of magazines, newspapers and television shows. Citizens groups, organized to combat Corps projects, are springing up at colleges and in many communities. At least four authors are writing books that will attack the Corps for allegedly disturbing the environment, in such a way as to make man himself an endangered species ultimately.

All this activity hasn't been lost on the Corps. Recently, the agency has added scores of biologists, ecologists and environmental scientists to its ranks. Their main assignment: Spot potentially embarrassing conservation problems before they can become public issues. The agency also has advised its regional officers to pay more attention to the environment and has refused to award some highly lucrative landfill permits that conservationists opposed.

"We wouldn't be candid if we didn't admit this new nationwide interest in conservation caught us pretty much by surprise." says Maj. Gen. Francis P. Koisch, a West Point graduate who has spent his entire 27-year career in the Corps. "We aren't deaf to this issue of preserving the environment. We're paying more attention now to the environment than we ever have in the past. We aren't too proud to admit we have made mistakes. But we also think our job is to mold the environment so it can be most effectively used by man."

Lt. Gen. Frederick J. Clarke, current Chief of Engineers, adds: "Meeting . . . demands for water must inevitably have an impact on the landscape. But we of the Corps believe . . . we can plan to meet the needs in ways that will result in a balanced and harmonious treatment of resources and environment."

Some observers feel the Corps may be getting more than its share of criticism. "The nation is really coming alive on the issue of saving our rapidly deteriorating environment," says a White House science adviser, "and the Corps of Engineers is the unfortunate cuss that has become the national scapegoat for all our environmental diseases." Such observers suggest that conservationists might do well to put more heat on the institution that authorizes all those controversial Corps projects: Congress.

Sen. William Proxmire, the Wisconsin Democrat, advocates sharply reduced budgets for the Corps. He calls the Corps "an oversized goose that lays thousands of golden pork barrel eggs a year" and adds: "Something must be done to put that big goose on a strict diet." Traditionally, Congress has been loath to impose the diet, because most Congressmen regard a new dam or canal as a handy aid to reelection.

But now there are signs of a changing Congressional mood. More than a dozen bills have been introduced that would force the Corps to pay more attention to environmental protection. The bills have support among both Democrats and Republicans and Congressional observers say several have excellent chances for passage. The powerful Senate Public Works Subcommittee, which reviews Corps appropriations, has said it will consider a measure that would make Corps projects much harder to justify economically.

Such a measure would limit the effectiveness of a key Corps tactic, the "cost-benefit ratio." In studying a potential dam project, for example, the Corps typically weighs the estimated dollar values of such expected benefits as flood control, recreation and perhaps hydroelectric power against the cost. For those projects the Corps decides to submit for Congressional approval, the benefits outweigh the cost—not surprisingly, say critics, who declare the whole procedure is rigged.

Critics often cite a study by Robert Haveman, a former University of Wisconsin economist who now works for a national resources organization in Washington, D.C. Mr. Haveman examined the economic benefits of 147 Corps projects between 1946 and 1962 that together cost almost $2.7 billion in Federal money. His conclusion: 63 of the projects, costing about $1.17 billion, "should not have been undertaken." Other economic studies have reached similar conclusions.

One bill before Congress would create a Department of Natural Resources that would take over many of the Corps' responsibilities. It was introduced by Sen.

Frank Moss (D., Utah) and is strongly opposed by the Corps.

Another bill, widely backed, is that of Rep. Henry Reuss (D., Wis.). He would give the Engineers statutory authority to build sewage treatment plants. "If the Corps is so eager to build things, then why not give it something worthwhile to build," Mr. Reuss reasons. The Corps backs this bill.

A third bill has already been passed by Congress and signed by President Nixon. It created a Presidential-level environmental council, to be composed of leading environmentalists appointed by the President. "The President could really do a job reforming the Corps through this council," says Rep. John Dingell, a Michigan Democrat. "Or he could make it just another one of this city's uncountable farces which we have to call executive agencies."

Regardless of the outcome of the bills still pending, Congress seems likely to turn new attention to the many controversial Corps projects currently somewhere in limbo between the early planning and actual construction stages. Among them:

—The Trinity River Barge Canal, which would link Fort Worth, Texas, with the Gulf of Mexico, 370 miles away. The Texas Congressional delegation has won Congressional authorization, but funds haven't yet been appropriated. The Corps says the project will cost nearly $1 billion and is economically justifiable, but several independent engineers and economists violently disagree. One suggests in jest that it would be cheaper to move Fort Worth to the Gulf.

—The proposed Salem Church Dam near Fredericksburg, Va. The Corps contends the dam would control floods, produce badly needed hydroelectric power, enhance "water quality control" and provide "recreational benefits."

Foes of the project say that it would flood one of the most scenic rivers in Virginia, the Rappahannock. They say the Fredericksburg area hasn't had a serious flood since 1946, that there are adequate recreational facilities in the area and that plenty of hydroelectric power is available from a publicly owned power company only a few miles away. With public protest growing, Interior Secretary Walter Hickel and several Congressmen have indicated they may try to block the project.

—The proposed $27 million Gaysville flood control project in Vermont. The project would destroy 110 homes and six public buildings, take 30 farms out of production and force relocation of one industry, says Vermont's lone Congressman, Republican Robert T. Stafford, who opposes it. Philip Hoff, who backed the project while Vermont governor, recently changed his mind and urged the incumbent governor, Deane C. Davis, to restudy the matter. Mr. Davis then requested that the Corps abandon the project. Corps officials say a governor's opposition usually is fatal to a project.

—A 120-mile-long canal that would link Lake Erie to the Ohio River, and also inundate nearly 90,000 acres of dairy farms and force relocation of more than 6,000 persons.

Also under attack is the Corps' operation of a network of canals above the Florida Everglades National Park. Park officials contend that in dry years the Corps diverts vitally needed water to farmers and that in wet years the Engineers dump the excess on the park. Everglades wildlife, dependent upon a steady supply of fresh water, is suffering badly, according to the park officials. Sen. Gaylord Nelson, the Wisconsin Democrat, calls the

Corps treatment of the Everglades "a national disgrace."

Such Congressional scrutiny is good news for the citizens groups that are opposing proposed Corps projects in about 30 communities. But it comes too late for many disheartened Floridians in Stuart, a town of 5,000 about 40 miles north of Palm Beach.

One resident, Dr. Frederick Krauskopf, still shudders when he recounts how one of his teen-age sons nearly drowned not too long ago when he jumped into the muddy St. Lucie River to clean the sides of the family boat. The boy landed in five feet of quicksand-like mud, and his struggles only made him sink deeper. Two brothers finally pulled him free. The incident badly upset many townspeople.

Col. John McElhenny, the Corps' district engineer for the Florida area, says he and his colleagues "felt badly when we heard about Dr. Krauskopf's boy." He says the Corps "is trying to do several things that will clean up the St. Lucie for sure, but it will take time." Aides estimate it will take 10 years—too long, in the opinion of many of the elderly retirees who make up much of Stuart's population.

"For Sale" signs have sprung up around the town. George Ewing, a 64-year-old retired stockbroker who has a home on the St. Lucie, says he has had four neighbors in the house next door in just two years. "One of my neighbors stayed for only five weeks," he says. "Finally, he packed up, moved out and told me: 'You can keep your ugly river and your ugly Corps of Engineers.' "

—TOM HERMAN

1970

Auto-Age Eyesores

BIG-CITY mayors must sometimes wonder whether Henry Ford did the world such a good turn when he came up with assembly lines capable of turning out cars by the millions. For one of the fastest-growing problems for many cities is abandoned cars that clutter streets in residential neighborhoods and blight the landscape along urban freeways.

"A derelict car hurts a street and a neighborhood in countless ways," says Mayor John V. Lindsay of New York. "It's the worst sort of eyesore. It attracts car strippers and vandals. It collects dirt and refuse. It becomes a dangerous playground for children. Narcotics addicts use it for a shooting gallery, and tramps use it as a flophouse."

Mr. Lindsay and community leaders in other cities across the country say the most troubling aspect of the "junker" problem is that so far it has defied solution. Moreover, the problem threatens to get much worse, for the scrap heap is growing faster than it can be removed or destroyed.

As recently as 1964, when 5.7 million cars were junked, all but 400,000 autos were consumed by industry for steel-making. In 1968, however, some two million car hulks remained to be disposed of after industry's needs had been met. In Mr. Lindsay's view, the growing junk pile is "a symbol of an affluent society and a wasteful society—a society that is barely able to control its own environment."

Many of the worthless junkers are abandoned at the curb in poor neighborhoods, sometimes by outsiders who bring jalopies in from other sections of town. That's apparent in an area such as Detroit's Fifth Precinct, a low-income, largely black neighborhood.

On many blocks there, two or three abandoned cars sit at curbside or in vacant lots. Often the wheels are missing and the windows smashed. Some are burned-out hulks. In one recent month, Detroit police hauled 1,437 abandoned cars out of the Fifth, up 37% from two years ago. Trying to keep up with the junkers "is like trying to sweep back water," says James Clarke, a Detroit police inspector.

Mr. Lindsay says that New York hauled away 21,500 abandoned cars during the first five months of 1969, compared with 30,000 during all of 1968. Mayor Alfonso Cervantes of St. Louis says: "A year ago, we rarely got a citizen's complaint about derelict cars, but in May they accounted for 22% of the 384 complaints my office received."

Several factors are behind the explosion in abandoned cars. There are more cars on the road to wear out. Motorists are also driving more; thus their cars wear out faster. The experts also say that cars depreciate faster these days than ever before, making their owners more likely to walk away from even a relatively late-model vehicle when it breaks down. Martin McInerney, a Detroit Chrysler-Plymouth dealer, says that 10 years ago "if you had a five or six-year-old car, you had a marketable product worth $250 or $300." Now, he adds, a five or six-year-old car is often worthless.

But the most important reason is that demand for scrap metal has been sluggish. As a result, demand for junkers has held level at about five million cars a year. But the seven million cars junked in 1968 were up 12%

from the previous year and almost 66% above 10 years ago.

A switch to steel-making processes that use less scrap is one of the main reasons the scrap metal market's consumption of junkers is no longer keeping pace with the pileup. Scrap steel from autos that brought $43 a ton in 1956 now fetches only about $20 a ton. Many scrap dealers say it costs them almost $20 to transport an old car hulk and then shred it into usable scrap, leaving a negligible profit. A few junk yard owners refuse to take old cars even when their owners offer them free.

The junkmen who do take cars are often highly selective. Julius Wainer, a Detroit scrap dealer, says he's only interested in junkers when he can make a profit selling individual parts, such as radiators. But like most dealers, Mr. Wainer only wants cars made after 1962. "The older ones aren't worth the trouble," he says.

In some cities, including New York and St. Louis, local governments must pay most of the cost of picking up junkers. Mayor Lindsay says that New York spent $585,000 last year to get private contractors to take away abandoned cars. St. Louis paid out $40,000 in the fiscal year ended last April 30 and expects to spend $100,000 in the current fiscal year.

Rising insurance and auto registration costs are also contributing to the junk-car problem. "By the time a guy pays for his plates and insurance, he's liable to wind up paying out more than his car is worth," says a Detroit policeman. And in Missouri, where a tough new safety inspection law recently went into effect, many drivers are junking their cars when they find out that the cost of repairs necessary to get their vehicle through the inspection line will be more than the car's market value.

Even if the streets could be swept clean of abandoned cars, that wouldn't solve the problem. Junk yard operators would still be left with ugly mountains of old cars, sometimes covering acres. Thus, ways are needed to dispose of the mountains.

Proler Steel Co. of Houston has built seven giant auto shredding machines, each capable of turning 250,000 cars a year into neat chunks of scrap. The machines are now operating in various parts of the country, but they are running at less than full capacity. The catch is that the low price for scrap metal holds down the amount Proler can pay for old car hulks, which makes junk yard operators less than eager to dispose of their old car stockpiles.

Maryland recently passed an auto disposal law that might make the use of such machines more feasible economically for scrap firms. The state will pay scrap processors $10 for each car they demolish. The money will come from small fees charged auto buyers.

The Maryland law also provides that junk yard operators will be allowed to keep cars that are seven model years old or older for a maximum of one year. After that, the wreckers face fines of $5 a car for each six-month period after the deadline that the car goes undestroyed. Several other states are considering similar legislation.

The Federal attack on the problem of disposing of old car hulks has been limited to research projects and to the moderately successful but scantily financed attempts to screen auto junk yards from public view. The screening program received no funds at all in the fiscal years 1969 and 1970. The Nixon Administration, however, says it will seek funds for screening in fiscal 1971.

Perhaps the most novel solution to the old-car problem comes from Warren B. Diederich, a Fargo,

N.D., contractor. He recently patented a method of crushing cars into interlocking shapes that could be coated with waterproof materials to prevent rusting. The idea is that the interlocking shapes would be a perfect substitute for crushed rock in building such things as breakwaters and jetties. But there is a considerable snag. The crushing and coating operation makes the method cost about 50% more than rock. Now Mr. Diederich is concentrating on reducing the cost of his technique.

—TIM METZ

1969

The Dying Lake

THE oil-slicked Cuyahoga River, which oozes its way through Cleveland to Lake Erie, catches fire periodically, earning it the dubious title of "the only body of water ever classified as a fire hazard."

The bacteria count of the chocolate-brown river water, which bubbles like a witch's brew because of fermenting gases on the bottom, often matches that of pure sewage. The river's mouth, once a wonderful spot to catch black bass, now is so contaminated that surprised scientists have found it doesn't even contain sludgeworms, which thrive on waste. Moreover, most local folks agree, the Cuyahoga stinks.

The Cuyahoga and dozens of other polluted Lake Erie tributaries are the focus of a historic and stormy struggle to save the fourth largest of the Great Lakes from becoming useless to man.

"Lake Erie represents the first large-scale warning that we are in danger of destroying the habitability of the earth," says Barry Commoner, a Washington University biologist and critic of current efforts to curb Erie pollution. "Mankind is in an environmental crisis, and Lake Erie constitutes the biggest warning."

Not all pollution experts will go that far, but there is general agreement on one thing: In little more than 50 years, man has wrought on the world's 12th largest lake a startling ecological change. And, biologists agree, the lake may be on the verge of a "biological explosion."

They also agree, however, that there's still a lot they don't know about Lake Erie. "We're only now beginning to understand the lake," says one biologist.

Experts figure the 9,940-square-mile lake, gouged out by glaciers 12,000 years ago, has been artificially "aged" 15,000 years in the past half century. The obvious by-products of pollution and aging abound. For 40 miles along the Cleveland area, Lake Erie waters, contaminated with slugs of raw sewage, human excrement, oil and chemicals, are generally unfit for swimming.

Algae, organisms invisible to the eye in a healthy lake, litter the Erie shoreline in long, rotting piles, clog city water intakes and add objectionable taste to many communities' drinking water. During summer, the algae collect in the western basin in an 800-square-mile mass two feet thick, turning the lake into a solution resembling pea soup.

Pollution also has killed most of the lake's commercial and game fish, such as northern pike, blue pike, sturgeon and cisco. Now the lake is populated by scavenger varieties, such as suckers and carp, which have little commercial value. In addition, the pollution is a significant factor in the sharp decline of wild ducks on Lake Erie and its tributaries. On the Detroit River, where in 1899 ducks were so plentiful one hunting club bagged 122 in a single hour, 10,000 ducks were destroyed by oil pollution in 1960 and huge numbers have shared similar fates each year since then, according to Federal wildlife experts.

The chemical interaction that has led to this massive devastation of Lake Erie was little understood until recently. Scientific information on lake ecology generally is at least 50 years behind that of rivers, according to one expert. And even now some aspects, particularly the effectiveness of recommended cures for Lake Erie, are controversial to say the least.

In the main, according to a five-year, $3 million Federal study, Lake Erie is choking to death on city sewage, although scientists prefer to call it over-enrichment or eutrophication. By far the biggest polluter, the study found, is the Detroit area, which contributes more troublesome municipal waste to Erie than all other U.S. cities combined. (Michigan denounces the study and is demanding sections referring to it be withdrawn.) The Cleveland-Cuyahoga River Basin ranked second. The lake is bounded by Michigan, Ohio, New York, Pennsylvania and Canada.

One obvious source of the pollution is raw, untreated sewage, piped into the lake as a result of inadequate or faulty sewer systems. For the last half of 1968 in Cleveland, for instance, a main sewer line was under repair and during that time up to 30 million gallons of raw sewage daily was channeled into the Cuyahoga River. The same situation prevailed for six months in 1967 while the line was down for repair. All told, the same line in Cleveland, whose residents recently voted $100 million to upgrade its antiquated sewer system, has had major breaks in each of the past four years.

Vast amounts of raw sewage also are channeled to the lake by combined storm-sanitary sewer systems, which during rains divert sewage through overflows. "Probably no city in the country has so many overflows, and each one is a source of pollution," states a city-financed study of the Cleveland sewer system.

For decades, pollution experts assumed such raw sewage was primarily responsible for lake pollution and ecological damage. As the organic material is broken down by bacteria, it uses up the water's oxygen, killing small and beneficial forms of animal life and even fish themselves.

But of late investigators have found that even greater devastation to the lake results from sewage plants, which had been considered modern and adequate. After treatment to remove indigestible solids and to break down organic material, sewage is discharged as about 85% inorganic products, particularly nitrate and phosphate. These consume no oxygen, and it was long assumed they were swept harmlessly on to eventually reach the sea.

Now the scientists believe that most of this phosphate and nitrate remains in the lake, where it fertilizes monstrous growths of algae, which bloom and quickly die. Although the phosphate and nitrate don't consume oxygen, the algae do use up enormous supplies of it. Total oxygen demand on Lake Erie from algae is placed at 18 times that of organic material from sewage plants.

As a result, nearly 25% of the bottom of Lake Erie is almost completely devoid of oxygen in certain periods, a situation that wreaks havoc on the ecology. Such life as nymphs of may flies, a choice food for such popular fish as bass and long considered the most common animal in Lake Erie, no longer exists in huge areas of the lake. Instead, sludgeworms abound at up to 30,000 per square yard. Similarly, fish that require high oxygen content, including pike, have been largely replaced by those requiring less, such as suckers and carp.

Worse, say biologists, the lake may be on the verge of a "biological cataclysm," involving sudden release of pollutants collected for a century in Lake Erie's bottom mud, which ranges from 30 to 125 feet thick. The result would be depletion of most Lake Erie oxygen, which could have an enormous additional impact on lake life. Even use of the lake as drinking water—it currently is used by some 11 million people—could be threatened, the biologists maintain.

The fuse to this biological bomb, scientists contend, is a thin skin of ferric iron overlaying the mud bottom of Lake Erie. With ample oxygen, ferric iron forms hard insoluble complexes with materials on the bottom, sealing off the deposits from the water above. But with depletion of oxygen ferric iron can become soluble, and materials in the mud then can dissolve in the water above. The release of enormous quantities of these materials, ranging from nutrients such as nitrate and phosphate to insecticides and industrial chemicals, has already begun, according to some experts.

Municipal waste, although Lake Erie's biggest problem, is not the only one. An industrial complex ranging from Detroit's auto plants and Cleveland's steel mills to Erie, Pa., paper operations and Akron rubber plants, constitutes the second biggest source of pollution. An estimated 11 million pounds of chlorides and undetermined amounts of cyanide, acids, oil and similar industrial pollutants are dumped into the lake every 24 hours.

The exact impact of these, particularly on reproduction of fish and birds, isn't known. But pollution experts say the danger is growing.

Despite the awesome proportions of the problem, many pollution experts are confident Lake Erie can be salvaged. "There's no question but that the lake can be saved, but I don't know that it will be," contends George Harlow, director of the Lake Erie office of the Federal Water Pollution Control Administration. "We can't fool around any longer."

The key to saving the lake, pollution experts figure, is removal from waste of most phosphorus "because it is the one essential nutrient for which control measures are best known." Some 137,000 pounds of phosphorus are dumped into the lake daily, with 72% of that stemming from municipal waste, studies show. About two-thirds of the phosphorus in city waste stems from detergents.

To remove phosphorus and cut down on untreated sewage, chemicals and other pollutants, the Federal Government is recommending immediate expenditure of $1.1 billion by cities and $285 million by industry.

But at least one biologist contends Lake Erie probably won't be saved even if all the steps recommended by the Federal Government are carefully followed.

"The idea of cleaning up Lake Erie by cutting back phosphates won't work," declares Mr. Commoner. "Nitrate levels also are serious." He adds further, "The job can't be done until we come to grips with the stuff on the bottom. To clean up Lake Erie may require the dimensions of the space program, and in the end it may even be irreversible."

—KENNETH G. SLOCUM

1969

Underground Dumps

WASTE disposal wells, conceived nearly two decades ago as a cheap way to get rid of industrial wastes without dumping them into streams and rivers, are turning out to be worrisome polluters after all.

That's the cry being raised by an increasing number of scientists and government officials as damage from the supposedly "safe" installations mounts. Consider these incidents:

—In Ludington, Mich., a leak in a Dow Chemical Co. disposal well polluted an underground water supply.

—In Erie, Pa., a well containing papermaking chemicals "blew its top," spewing thousands of gallons of the foul-smelling liquid into nearby Lake Erie.

—In Denver the pressure caused by the Army's pumping of poison gas wastes underground is blamed for causing a spate of earthquakes that still are going on even though the pumping has ceased.

Worry over disposal wells is mounting with their use. Faced with stringent new antipollution laws designed to protect the environment, more and more companies are turning to wells to get rid of liquid wastes. The companies contend that when properly installed and operated, the wells can be the most effective means of disposal. But some others aren't so sure.

"Underground waste disposal is a potential environmental Frankenstein's monster," declares Secretary of the Interior Walter Hickel. "All prudence dictates that we attack this problem systematically before it gets out of hand."

The number of disposal wells isn't believed to be large as yet. One recent survey puts the figure at 150 in all. But some authorities believe the total already is much higher because hundreds of wells were drilled before many states required permits and more have been drilled illegally since. The present total compares with only 30 wells a decade ago, and experts predict the number will grow rapidly in the future.

Disposing of wastes underground is basically simple. A hole five or six inches in diameter is drilled into the ground until it reaches a layer of standstone or limestone, usually 2,000 to 3,000 feet down. These porous formations act like a petrified sponge, soaking up liquid wastes that are pumped into them. The hole is encased in steel so wastes won't seep out the sides into drinking-water supplies that lie along the way.

Such a well can absorb huge quantities of waste. Pumping sewage at the above-average rate of 500 gallons a minute, it would take 200 years to saturate a 100-foot-thick layer of sandstone covering only 20 square miles. This is a relatively small formation; some measure 2,000 feet thick and cover 20,000 square miles.

A wide variety of wastes is being dumped into such underground "reservoirs." U.S. Steel Corp. puts hydrochloric and sulphuric acids left over from steel-treating into a 4,300-foot well at its Gary, Ind., works at the rate of 6.5 million gallons a month. Velsicol Chemical Corp., a subsidiary of Northwest Industries, uses a well in southeastern Illinois to get rid of 60,000 gallons of poisonous residues from pesticide manufacture daily. International Salt Co. puts waste brine from salt-mining down a well near Elmira, N.Y., and DuPont Co. disposes of chemicals from Teflon production in West Virginia the same way. A self-service laundry in Oakland, Mich., even dumps its dirty wash water into a well.

The method appeals to industry at least partly because it normally involves a small initial cost, usually around $100,000, though some wells can be dug for as little as $20,000. By comparison, sewage treatment plants with settling tanks and other equipment for cleaning up wastes before they are dumped into a river typically cost $500,000 or more.

Wells are cheap to operate, too. Standard Oil Co. of Ohio puts waste from production of acrylonitrile, a flammable liquid used in making textiles and plastics, down a well. It figures it saves $500,000 a year by disposing of the material this way instead of burning it.

The wells' threat to the environment, however, outweighs the economic advantages, some scientists say. One big worry is where the wastes go once they get underground. The fear is that they could escape and contaminate underground water or mineral deposits over thousands of square miles.

Companies say they are careful to inject sewage only into sandstone or limestone formations that are sandwiched between impermeable layers of shale that prevent wastes from escaping. But critics say companies also like to engage in a practice called "fracturing," in which the wastes are pumped into wells under very high pressure. This cracks the rocks in the storage formation, thereby increasing a well's capacity, but it also can crack the shale, allowing wastes to seep out of the disposal "reservoir."

"We're just lucky we haven't had any major underground contamination so far, because the potential is there," warns E. L. Hendricks, chief of the U.S. Geological Survey's water resources division.

There already has been some underground pollution. In late 1969 the casing of a 2,500-foot-deep well operated at Ludington, Mich., by Dow Chemical Co. corroded at the 450-foot level, spilling thousands of gallons of brine into an underground freshwater pool. The exact amount of the spill is unknown because no one knows when the leak developed. Dow says it thinks the well leaked only two or three days before it was shut down but it concedes it could have leaked as long as three months. Fortunately the pool of underground water has not been tapped as a source of drinking water. The repaired disposal well was put back in service a few months later.

Disposal wells can pollute surface waters, too. Two years ago one of three wells operated by Hammermill Paper Co. in Erie, Pa., sprang a leak. The waste—a brown, putrid broth left over from processing wood chips into pulp—had been forced into the well under high pressure. When the trouble developed, the compressed liquid shot to the surface, forming a 20-foot geyser that spilled into nearby Lake Erie. It took three weeks to cap the well. By that time, more than two million gallons had escaped. In six years Hammermill has pumped 750 million gallons of the broth underground.

A different pollution problem has arisen in Port Huron, Mich. Officials of the Michigan Natural Resources Department say that chemical waste dumped into wells in Sarnia, Ontario, across the Detroit River from Port Huron, has caused a buildup in general underground pressures. This, in turn, has caused nearly a dozen old, abandoned oil and gas wells beneath the city of Port Huron to begin leaking.

In one case underground pressure forced crude oil up through an old well and through a concrete parking lot at the city post office. The parking lot had to be torn up and resurfaced and the well located and plugged.

In another case, oil seeps up through the basement floor of one Port Huron home at the rate of several gallons a month. Efforts to locate the leaking well and plug it have failed so far. In still another instance, flammable natural gas was found escaping into the air from a well only a few feet from a local hospital.

The disposal wells allegedly causing the trouble are owned by several oil companies, including Sun, Imperial and Shell, that operate refineries nearby. State and company officials have failed to agree on how to solve the problem. Michigan asked the oil companies to stop using the wells, but the companies contend that there's no conclusive evidence that their wells are causing the leaks in Port Huron.

Nonetheless, at least one company, Imperial Oil Ltd., says it plans to abandon the five disposal wells at its Sarnia refinery and instead will put the waste down a new well being drilled 20 miles south. According to the company, the new well definitely won't contribute to the problem.

Companies generally defend underground disposal by saying it gets waste completely out of the human environment. They also say that the wells comply with state regulations. A few companies say that they make extensive geological studies to determine the suitability of underground rock for waste injection before they drill a well and that they construct wells of corrosion-resistant materials that won't leak.

Such safety procedures can be expensive; Reichhold Chemicals Inc. is spending $675,000 on a well at its Tuscaloosa, Ala., plant. There's evidence, however, that most companies aren't willing to spend the sums needed for proper installation of the wells since this would destroy the economic edge the method provides, critics say. As a compliance with state regulations, critics point out that only a handful of states have laws specifically regulating underground disposal.

Aside from the pollution peril, many scientists say that pouring wastes underground is dangerous because the subsurface is the least-understood part of the earth and no one really knows how the wastes will behave.

Florida water pollution officials, for instance, frankly admit they don't understand why a well in Orlando, operated by a citrus fruit processor, erupts every so often, spitting diluted orange juice and other citrus wastes onto company grounds. "It may have something to do with the flow of an underground river," says one official. The company, Southern Fruit Distributors Inc., which refuses to comment on the matter, is planning to abandon the well in favor of a surface waste treatment facility now under construction, according to state officials.

Scientists also say that the wells' impact on subsurface structures is dangerously unpredictable. They claim, for example, that in more than one case wells apparently have triggered earthquakes. In 1962 the Army drilled a 12,000-foot well at its Rocky Mountain Arsenal in Denver to dispose of waste from the manufacture of poison gas. A month after injection began Denver had its first earthquake in 80 years, and in the next five years the area experienced more than 1,500 quakes. No buildings have collapsed, but cracked windows and foundations have been widespread.

"Periods of increased pumping resulted in many quakes and periods of decreased pumping resulted in fewer quakes," says David M. Evans, a Colorado School of Mines geologist who studied the situation. "All the earthquakes originated near the well."

After pumping 150 million gallons of poisons into the hole, the Army abandoned it in early 1966. Since then the number of quakes has diminished. What probably happened, Mr. Evans says, is that the pressure of the injected wastes forced open cracks in the underground rock formations far enough to start the rocks sliding and shifting, producing the earth tremors.

Recognizing the problems arising from disposal wells, states are beginning to tighten their regulations. Illinois no longer allows wells to be drilled in a 2,500-square mile area around Chicago. This is to protect an underlying reservoir of slightly salty water that may be a future source of drinking water.

New York state recently declared that it regards underground disposal as a "last resort" after all other methods have been tried. Florida has halted all drilling of the wells until it can revise its regulations.

The Federal Government is taking more interest, too. Interior Secretary Hickel has ordered the Geological Survey to take the lead in a research program to evaluate the effects of underground waste disposal. "There may be no perfectly safe way of disposing of waste underground," he said. "But we must review existing regulations and start collecting the kind of environmental data needed to assess the level of risk and consider ways of organizing Federal, state and industrial efforts to solve this growing problem."

—RICHARD D. JAMES

1970

Unwelcome Neighbors

THE nation's electric utilities feel like outcasts. Everywhere they go, somebody wants to run them off.

Time was when most communities were only too glad to hear that utilities planned to build power generating plants in their areas. A plant meant jobs and it usually meant a huge new addition to the tax rolls of the community—resulting, often, in a sizable property tax break for homeowners.

But times have changed. With the new surge in concern over the environment, the smoke, water pollution and general unsightliness typically associated with power plants have made them unwelcome neighbors in many places. More and more utilities are finding that it's all but impossible to find a new power plant site that doesn't stir alarm among local residents and conservationists.

Southern California Edison Co., for instance, has been rejected by two towns in which it wanted to build new electrical generating capacity. The first was Victorville, a small town on the edge of the Mojave Desert 90 miles northeast of Los Angeles. Southern Cal Edison had planned a coal-fired generating plant there but abandoned the idea after local citizens protested strenuously. "People just didn't want that thing in their back yards. It became a very emotional issue," says James Cox, Victorville's city manager.

Southern Cal Edison then began planning an expansion of its Huntington Beach plant in Orange County, within the smog-ridden greater Los Angeles basin. Again it was stopped cold. County authorities refused to issue a construction and operating permit for the expansion, arguing it would add to pollution.

So, for now at least, the company is contenting itself with expanding an existing nuclear power plant at San Onofre, some 60 miles south of Los Angeles.

But company officials say that they must build new plants somewhere in Southern California before long to meet the growing power needs of the area. In so doing, they echo other utility executives throughout the nation who warn of impending blackouts and brownouts because of stalled projects.

Conservationists and those concerned with aesthetics retort, however, that a livable environment is worth the inconvenience, and they continue to fight plants they regard as objectionable.

The conflict couldn't come at a worse time for power suppliers, who are scrambling to meet consumer demand expected to nearly double every 10 years. The Federal Power Commission estimates that U.S. electrical consumption will reach 5.8 trillion kilowatt hours annually by 1990, compared with the 1.5 trillion kilowatt hours required in 1970.

No widespread power shortages are foreseen during the next decade, but the FPC says a number of individual utilities already have dangerously slim generating reserves—capacity above that needed in routine peak periods. Thirty-nine of the 181 major power systems in the country have reserves of less than 10%, compared with the 15% to 20% considered adequate by the FPC.

New York's Consolidated Edison Co. already has warned that it may be forced to black out residential areas of New York City for short periods on a rotating basis to avoid widespread disruption of summertime service during peak periods when air conditioners are in heavy use.

The company, the target of criticism from many quarters for years for alleged inefficiency and insensitivity to consumer needs, is now 10 years behind schedule on a planned generating plant at Storm King Mountain on the Hudson River north of New York City.

The principal cause of the delay has been the Scenic Hudson Preservation Conference, a body set up in 1961 specifically to fight the project. The group claims 20,000 members. It has dragged Con Ed through the courts and has sent its lawyers and representatives to hearing after hearing on the case, trying to force the company to find alternative means of providing power.

The company says it has been forced by the Storm King delay and by delays in construction of nuclear power plants to expand a coal and oil-burning plant in the congested Astoria section of Queens in New York City.

"That's as phony as a $14 bill," claims Rod Vandivert, executive director of Scenic Hudson. He contends that Con Ed's management has deliberately delayed finding other alternatives to increase the pressure for a plant at Storm King—an allegation a Con Ed spokesman says is "absolutely false."

In Los Angeles, a power plant battle has erupted between two governmental agencies—the Los Angeles Department of Water and Power and local air pollution control authorities. Like Southern California Edison, the city agency was rebuffed in an attempt to build a power plant at Victorville. It then spent $6.5 million expanding its existing coal-fired Scattergood plant near Los Angeles—only to be stalled by a pollution control board order forbidding use of the new generating unit.

An official of the power agency admits that construction was started without full clearance from pollution authorities. "The board always came through in the past, but it sure didn't this time," he says.

Floyd L. Goss, chief engineer of the power agency, says the ruling "could easily result in a partial blackout of the city by 1972." An earlier attempt to build a nuclear plant at Malibu, just west of Los Angeles, was stalled by public fears over earthquake hazards and by eventual refusal of the Atomic Energy Commission to grant a permit.

In Minneapolis, Northern States Power Co. is facing a crisis over a nuclear reactor it has built and is testing on the Mississippi River north of the city. A Mother's Day march on the governor's mansion in St. Paul in 1969 dramatized citizen protests against the facility, and two environmentalist groups contend its radioactive emissons could cause birth defects and leukemia.

State authorities have ruled that the $90 million plant can't be operated unless its radioactive emissions are cut to one-fifth the level allowed by the Atomic Energy Commission. "Who are we supposed to answer to—the AEC or the state?" asks a spokesman for the company. He complains that the state standards are "technically impossible to meet" and warns of a "critical power shortage" in the area during summer peak-use periods.

In some areas, power companies are placating conservationists by spending heavily on antipollution equipment and efforts to make power plants look better. Commonwealth Edison Co. in Chicago, under fire from public and private groups as a polluter, recently announced it was cutting back sharply on its coal buying and switching to low-sulphur oil that produces far less pollution. But officials said the move would hold the company's profit gain in 1970 to no more than 5%. Earlier they had expected a gain of as much as 8% over 1969 earnings.

In Ohio, three utilities will be spending $22 million more than originally expected to combat thermal pollution of the Ohio River near the town of Moscow, where they are planning two jointly owned nuclear generating units. The money will go for cooling towers, which will discharge excess heat into the air rather than into the water, where it would affect plant and animal life.

"Beautification" of transmission lines and substations—through artful landscaping, more attention to aesthetics in design and, in some cases, the use of underground lines—is increasing. "You can't tell some of these stations from the houses next door," boasts one utility executive.

The extra costs of environmental protection haven't yet been reflected in any widespread major increase in electrical rates. But power suppliers feel that rates definitely are headed up. An official of the Salt River Project, a big Arizona-based irrigation and electric cooperative, says that "a turnaround in the long-term downward trend in the cost of power" is inevitable in view of heavy spending on environmental protection.

Conservationists say they don't mind paying a little extra if the air and water are made cleaner. But they do fault the utilities for their heavy promotion of "all-electric living" at a time when, in the conservationists' view, the need to protect the environment requires that only the most necessary plants should be built.

The Sierra Club, a conservation group whose membership jumped from 70,000 to 86,000 in 1969, is supporting a bill by Sen. Edmund Muskie of Maine that requires regional planning of power plants. The intention would be to discourage construction of countless small plants in favor of fewer big ones, with controlled emissions, to serve regional needs.

The club already has been involved in lawsuits against power suppliers and threatens more. For example, if the FPC approves the proposed Hell's Canyon Dam on the Oregon-Idaho border, a project backed by a consortium of power suppliers in the Pacific Northwest, the club will go to court to stop it, a club official says.

In the meantime, battle lines may be forming in the thinly populated Southwest, where a group of utilities and other suppliers have plans for big coal-burning plants. Desert dwellers are already uneasy over pollution, complaining that 250 tons of pollutants are thrown into the air daily by the Four Corners generating plant in Northwest New Mexico. The facility, built in 1962 by Arizona Public Service Co., is called "the school of horrible examples" by another utility executive.

The New Mexico Citizens for Clean Air and Water complain that fly ash from the plant has limited visibility sharply Los Alamos physicist Joseph Devaney, who takes a light plane aloft periodically to check on the spreading haze, says: "I admit to a certain disbelief one day last fall when I couldn't even see the colors of the Grand Canyon." He blames particles drifting westward from Four Corners.

Now two more units are being installed at Four Corners for six Southwestern power suppliers, principally Southern Cal Edison. Another major plant, the Mohave, is under construction near Bullhead City, Ariz., and two gigantic plants are planned near the Lake Powell area in northern Arizona, one near Page, on the Navajo reservation, and one on the nearby Kiaparowitz Plateau.

The utilities are taking special pains to soothe public opinion. Arizona Public Service says it will spend $7 million on its own existing units at Four Corners to capture fly ash. The two new units under construction will have their own precipitators. Even when the expanded plant is in full operation, total emissions will be about a third of what they are today, companies say.

Many conservationists remain unconvinced. They fear that the huge plants going in or planned may pollute a big chunk of the Southwest by the late 1970s, including the Navajo reservation and extensive parklands —even with the latest antipollution gear built in.

—BILL SLUIS

1970

The Mercury Threat

HEADACHE, fatigue and irritability. Those are the initial symptoms of mercury poisoning, though the cause isn't likely to be known when they appear. Trembling hands, slurred speech, numbness of the limbs, kidney damage and even death may follow in severe cases.

Therein lies a potentially explosive public health problem. Alarmed health officials in the U.S. are increasingly finding mercury in the nation's food supply. Within the past few months dangerous quantities of the chemical have been detected in hogs in New Mexico, cattle in Oregon, pheasants in Montana and California and fish in the Great Lakes.

Experts in the field assert that mercury isn't just another pollution problem. Whereas the specific effects of auto pollutants, insecticides and pesticides on the human system aren't certain, for instance, mercury in sufficient quantity is a lethal poison for which there is no known antidote or cure. Even mercury vapor is harmful.

Moreover, human beings and animals tend to retain mercury in their systems. Minute amounts may accumulate, becoming apparent only when irreversible damage has been done. "Mercury in virtually any quantity will cause some damage to people, but the effects of small amounts won't show up until you get older," says Dr. Alan R. Hinman, an expert on diseases of the brain and the nervous system at the National Communicable Disease Center (NCDC) in Atlanta.

It's known that about 10% of any mercury ingested goes to the brain. "We have lots of brain cells," says Dr. Hinman. "We may lose only a few with each tiny amount of mercury, and the impact might not be noticeable until we're older. Possibly some persons considered senile are suffering from long-term mercury poisoning."

There are two sources of mercury contamination. One is agricultural. Fungicides containing mercury are applied to the seeds for most grain crops, such as wheat, barley, oats and rice, and also to seeds for tomatoes and cotton, before the seeds are planted.

The purpose is to prevent diseases that could afflict these crops at early stages. The treated seed is dyed red as a warning. But, say officials, many farmers ignore the warning and feed leftover or waste seed grain to their animals. The animals pass the mercury to human beings who eat the animal meat. Game birds such as pheasants accumulate the poison by eating seed grain before the seeds sprout.

The second source of pollution is industrial. Waste containing mercury results from processes for making plastics, chlorine and sodium. This waste, discharged into lakes or rivers, accumulates in fish later taken for food.

Alarming information on mercury poisoning has reached U.S. investigators from abroad. Between 1953 and 1960, more than 60 persons were poisoned fatally by mercury wastes discharged into a river in Minamata, Japan, by a plastics manufacturer. Poisoning came from fish and shellfish caught in the river and bay and eaten.

Dr. William H. Likosky, a physician at the Atlanta

NCDC, says, "We have received reports that people in Guatemala, Pakistan and Iraq have developed illnesses through misuse of seed grain as food. They apparently have ground the grain into flour and made bread. There are hundreds of cases we know about, and there probably are many more that go unrecognized."

Federal officials don't know how widespread mercury poisoning is in the U.S. "These abuses have been around for years," says Dr. Likosky, "but no one added them together into a big picture. If someone had a mild case of mercury poisoning, it probably would have been diagnosed as encephalitis (a brain infection) of unknown etiology (cause). We have no idea whatsoever how many cases of mild mercury poisoning are occurring."

Concern mounted in the U.S. in late 1969, however, after three children in the family of Ernest Huckleby, a school janitor in Alamagordo, N.M., became ill. U.S. Public Health officials learned that the family had fed their hogs waste seed grain that had been treated with mercury compounds. Then the family slaughtered and ate one of the hogs.

The three children remain seriously ill. Investigators learned that the Alamagordo problem didn't end with the Huckleby family. Four other farm families in the region had obtained contaminated waste seed grain and possibly fed it to their animals. The U.S. Department of Agriculture clamped an embargo on hog shipments from the region in January 1970—but not before 108 possibly poisoned hogs had been slaughtered and shipped out. Federal officials say they think most of the pork went to New Mexico and Texas.

A serious problem is detecting contaminated meat. Farm animals fed mercury-treated seed aren't likely to be intercepted by inspectors from the USDA, concedes

Dr. John Spaulding, head of the toxicology group in the USDA consumer protection program. Federal inspectors check animals before they are slaughtered, and "only go on how the animals look," he says. Seven contaminated cattle that had been passed by a Federal inspector in Oregon were intercepted only because a tipster angry at the rancher notified the USDA.

Another problem is how to prevent farmers from using the contaminated grain. Many seed grain companies take their waste mercury-treated grain and leave it at dumps. But officials report at least one recent case of such grain being picked up and resold as feed grain. Other seed companies burn their grain. But this spreads dangerous mercury vapor about the countryside.

Early in 1970 the USDA took a strong but little-noticed move by suspending Federal registration for an estimated 60% of the mercury compounds used to treat seed grain. "Directions for proper use and caution statements on labels of the product have failed to prevent its misuse as a livestock feed," says a spokesman from the USDA's agricultural research service.

In order not to hamper planting, the department is allowing seed processors to sell seed already treated with mercury. However, the USDA indicates that it may eventually extend the partial ban and withdraw registrations for all mercury seed dressings.

The agricultural industry, however, complains that the Government is moving too swiftly and forcefully against mercury fungicides. Denis Hayley, secretary-treasurer of the National Agricultural Chemicals Association (NACA), a Washington trade group, estimates that about 80% of all commercial seed, including cotton, is treated with mercury compounds.

"There's nothing I know of with the broad spectrum of use against disease that could replace mercurial compounds," Mr. Hayley says. "The other available treatments are generally specific for a certain type of disease of a seed or for a certain crop."

Cutting off the mercury compounds, Mr. Hayley says, "will raise the cost of agricultural production substantially." Parke Brinkley, president of the NACA, asserts that "some 20% of the grain crop or yield would be lost without seed treatment, and that would be disastrous in this country." At least one fungicide maker already has challenged the partial Government ban in court.

The contamination of fish and other marine sources of food is an equally critical problem. "We are finding fish from the Great Lakes with mercury residues," says R. E. Duggan, a deputy associate commissioner of the Food and Drug Administration. "Just how widespread this is we don't know."

The FDA now has dozens of investigators checking fish in the Great Lakes and in several major rivers, including the Mississippi and the Niagara. Shrimp and oysters harvested from the Gulf of Mexico also are being checked. The FDA has set a figure of 0.5 parts per million as indicating contamination, and Mr. Duggan says "a substantial number" of contaminated fish have been found. But the average contamination among fish in lots checked has been below that level.

The Government doesn't have the power to halt fishing. But it could ban interstate shipments of contaminated fish or shellfish, and Mr. Duggan says, "We will take legal action against any fish shipped in interstate commerce above that level (0.5 parts per million)."

Canada and the states of Ohio and Michigan, armed with more direct powers, already have begun to use them. Paul W. Brown, attorney general of Ohio, has announced the preparation of legal action against Dow Chemical Co. and Wyandotte Chemical Co. to recover damages for alleged pollution by the companies' chlorine-caustic plants.

Dow acknowledges that its plant, located in Sarnia, Ontario, was pouring up to 200 pounds of mercury a day into the St. Clair River, which runs into Lake St. Clair, the Detroit River and Lake Erie. But it says pollution from the plant has been reduced to "essentially zero." Canada, however, has banned all fishing from its side of the St. Clair River and the Detroit River and in Lake St. Clair. Ohio has banned commercial fishing for some kinds of fish in its portion of Lake Erie.

Wyandotte is a subsidiary of Badische Anilin & Soda-Fabrik AG, a German company. Its plant, located on the Detroit River south of Detroit and two miles north of Lake Erie, had been emitting 10 to 20 pounds of mercury waste a day into the river. Michigan Attorney General Frank J. Kelley got an injunction that closed the plant for a few days until the company eliminated the pollution.

Another area of concern is catfish grown commercially on Southern rice farms, a growing industry. "They've been using mercury fungicides in the rice belt for 20 to 30 years," says William Barthel, chief of the FDA's toxicology laboratory in Atlanta. "I'm not concerned so much about the rice. The question is, has it (the mercury) built up to dangerous levels in the water?"

The FDA is buying a new tool to help in the Government's mercury-hunting effort—an atomic absorption spectrophotometer that will enable scientists to detect mercury in amounts as minute as 1-100th part per million in blood or urine specimens.

There is other work on the research front. Some

Swedish studies have suggested the existence of a process called translocation, in which mercury in seed dressings apparently passes from the seed to the resulting plant. U.S. experts are checking the Swedish conclusion. Sweden has banned the use of most mercury fungicides.

An unusual program is planned by the National Communicable Disease Center in Atlanta. Mercury poisoning falls outside its usual purview, but it plans to use its resources to check human beings and livestock around the country to determine how widespread mercury poisoning is.

—RICHARD R. LEGER

1970

Our Garbage Glut

WITH 100,000 tons of uncollected garbage piled in the streets, New York City captured the limelight during its celebrated strike of sanitationmen. But knowledgeable insiders in the trash field weren't impressed. Compared with what the future threatens nationwide, they say, this was a case of petty litter.

The headlong rush of the Affluent Society toward ever greater abundance apparently is generating so much garbage that it soon won't be possible to dispose of it without major technological innovations. "We're running in front of an avalanche, and it's already begun to bury us," warns Karl Wolf, a Chicago specialist.

Although the 1968 strike of New York sanitationmen caused a furor, day in and day out the average citizen around the country doesn't consider garbage a crucial problem, and it tends to get low priority in municipal budgets.

But the experts have compiled alarming statistics. One says the 7 million tons of refuse New York City produces annually would form a mile-high mountain in Yankee Stadium. Whereas the average American was responsible for 2.7 pounds of trash per day in 1920, the figure now is more than 4.5 pounds, and it is expected to reach 6 pounds within 10 years. Another curiosity: It costs more to dispose of the huge Sunday New York Times than the reader pays for it, one study claims.

Existing systems for getting rid of trash are largely obsolete or inadequate. But waste disposal visionaries are toying with some ingenious new ideas. One is to compress garbage into bricks about one tenth the original volume, using heat and pressure to create a stable product that could be used for land fill or, when sheathed, as a building material. Another suggestion: Compost refuse into fertile humus for crops. Some city officials advocate transporting urban trash to rural areas to fill abandoned mining cavities. In DuPage County, Ill., near Chicago, a ski slope 188 feet high is being constructed by heaping up garbage and dirt and compacting it all with bulldozers.

The situation is challenging. In a study, one packaging trade association found that a typical mile of highway yielded three cubic feet of trash, including 2,665 items ranging from papers to bottles to pajamas. The association said it costs about 32 cents an item for disposal. A Presidential Science Advisory Committee reported that the nation annually must dispose of 48 billion cans, 25 billion bottles and jars and 65 billion caps and tops.

A disconcerting thought for the future is the proliferation of "throwaway" products, from dresses and diapers to furniture and even houses (proposed by some planners as interim dwellings for ghetto residents between demolition of tenements and construction of new housing). And many of the new "disposables" aren't destructible; plastic and aluminum containers are examples.

"Convenience packaging, gentlemen, is rapidly becoming problem packaging, and we are all caught right in the middle," William F. May, chairman of American Can Co., told packaging executives at a conference.

Big cities are most troubled by the disposal problem. Incineration, a standard method, tends to create unhealthful air pollution while it disposes of garbage. At various times, municipal incinerators in New York City, which spew 17,233 tons of soot into the air yearly, have been shut down for several days so soot could dissipate.

"The air pollution people tell me, 'Don't burn it, dump it,' " says Maurice Feldman, New York's acting Sanitation Commissioner. "And the conservationists cry, 'Don't dump it there, that's a bird breeding area.' We're trying to figure out what to do with it."

Commissioner Feldman, an engineer who says, "I've been bouncing back and forth from garbage to sewage all my life," is embattled. He has investigated various new ideas for garbage disposal, but, he says, "No one's yet shown me the ideal solution."

Sanitary land filling, one of the most common disposal procedures, has consumed nearly all the accessible open acreage around many major cities. Within five to ten years, the experts predict, a large city without a shortage of land-fill area will be a rarity.

Mr. Feldman says New York's land-fill space will run out in four to nine years. A new incinerator or other mechanical facility is in the offing, but that will take six or seven years to complete. In the meantime, the city will squeeze trash into odd corners of existing land-fill sites to get by. "It's a neck-and-neck race," Mr. Feldman says. "If we fall behind, it'll be a catastrophe."

A sanitation official's nightmare came to pass during New York City's nine-day strike of sanitationmen, dramatizing some possible problems of the future. Accumulating at the rate of 10,000 tons a day, the trash formed head-high piles in some tenement sections and even disfigured the swank precincts of Park Avenue.

"At least this will help people see the size of the job

we have," Mr. Feldman sighed during the height of that crisis. Had it been summer, when food waste would have rotted quickly, stench and rats would have made many sections of the city unlivable. As it was, a health emergency was declared, and preparations were made for mass immunizations against typhoid fever. (One desperate New York husband churned up food waste in his electric blender and then flushed it down the toilet.)

New York's land-fill program uses up 200 acres of swampy lowland a year. Chicago faces a problem comparable to New York's. A giant land fill on Chicago's South Side is full, and planners are scrambling to find other means of disposal.

Bids solicited by Chicago on an ultramodern incinerator came in at $11 million, $4.5 million above the original estimates, and the city rejected them. One major source of the prospective increased cost, according to Streets and Sanitation Commissioner James V. Fitzpatrick, is high-efficiency electrostatic precipitators that take particles out of incinerator smoke. Distasteful as this kind of expense may be, the experts say municipalities are going to have to bear it.

The garbage problem also has reached crisis proportions in San Francisco. For decades, it and surrounding communities dumped waste in marshlands situated along the edge of San Francisco Bay. But the once-blue bay turned an alarming olive-brown, and bay filling was banned several years ago. Among those protesting were conservationists who said the natural habitat of various birds and fish was being destroyed. San Francisco now is considering an elaborate incineration process in which steam produced by the burning might be sold to a utility.

The situation is getting equally sticky in San Mateo County outside San Francisco. A while back the county settled on a new place to dump waste—Green Valley, a seaside canyon only a stone's throw from Coast Highway 1, recently designated a National Scenic Highway. Despite a hue and cry from conservationists, the county supervisors approved the site.

The story is the same around the nation. "Every large community is in a corner where it must do something about the garbage that threatens its very environment," says Mr. Wolf of Chicago, who is director of research on waste disposal for the American Public Works Association. "America is at the crossroads where it could do irreparable harm to itself."

Experts say a lasting solution is going to require cooperation between urban and rural areas. One notion being considered by Chicago planners involves a dust-to-dust concept. A city's waste comes from minerals, wood and other materials produced in rural areas, they say, so why not ship waste back to the hinterlands by rail and deposit it in scarred mining lands?

According to Mr. Wolf and other planners, the "rail-haul" system would have many benefits. It would aid railroads by employing entire freight trains to tote trash daily. It would be a boon to the cities. And it would help reclaim much of the land made useless by strip mining (the U.S. Bureau of Mines says strip mining takes about 100,000 acres out of use each year).

This notion also is being talked about in New York and Philadelphia. But rural sections that would be on the receiving end of this largess are bristling. Mr. Wolf concedes that an intense "education program" would be necessary to persuade the rural people that the garbage will assault neither eye nor nose.

Some experts are suggesting a refinement of the rail-haul system: Crush the waste first with a compacting device. While this shrinks waste no more than incineration—to about one tenth the original volume—the compacting machine requires an initial investment only a fifth as large as an incinerator.

Commissioner Fitzpatrick of Chicago traveled to Tokyo to inspect a compacting machine made by Tezuka Kasan Co. (a "junk junket," political opponents on the city council charged), and the city decided to invest $1.5 million in a pilot project. One problem: What to do with potentially germ-breeding ooze squeezed out of the garbage in the mashing process.

Tezuka Kasan is using such a machine to produce building blocks from crushed garbage. The blocks are sheathed in steel.

The garbage specialists are mindful of a classic tenet of science, the Law of Conservation of Matter. While it is possible to change the nature of a substance, the law states, it isn't possible to make it disappear entirely. Thus, the ultimate solution for garbage would seem to be changing it into something useful.

Industry has begun to explore the challenge. A British engineering firm announced that it had developed a chemical composing process that converts all sorts of garbage ("even the kitchen sink") into a liquid peat for fertilizing crops.

Other companies are exploring similar processes. But such fertilizer isn't as rich in plant nutrients as inexpensive chemical fertilizers, and commercial demand for it has proved small. And there is another problem: One such plant, described as a "showcase," was shut down in Florida because it produced a horrific stench.

But, with garbage disposal and collection costs at $3 billion a year nationwide, and rising, Mr. Wolf is convinced that industry "will find a very fertile field for research and development" in trash.

—ALAN ADELSON

1968

Radiation: How Much Is Safe?

LIKE ancient Troy, the U.S. Atomic Energy Commission is surprised to find an enemy within. The resulting uproar threatens to scramble industry and Government plans for expanded uses of atomic energy.

The Trojan horse, in the eyes of angry AEC officials, comprises a pair of well-known scientists at the famed Lawrence Radiation Laboratory in Livermore, Calif., an AEC-financed facility run by the University of California. The men, Dr. John Gofman, a medical doctor and nuclear chemist, and biophysicist Arthur Tamplin, charge that the AEC is risking "genocide" by approving dangerous levels of radioactivity for peaceful atomic energy uses.

The outcome of the scientists' attack remains in doubt, but its implications are profound. From the AEC and nuclear industry standpoint, the controversy could upset plans for vast development of nuclear electric plants and the unlocking of billions or even trillions of dollars in natural gas, shale oil and other mineral riches through "Plowshare" underground nuclear blasts.

The two Lawrence scientists say that their concern is public health. They charge that if Americans receive the maximum "safe" or permissible dose of radiation set by the Federal Radiation Council and endorsed by the AEC, there may be 32,000 extra cancer deaths yearly.

Their findings are based on statistical analysis of world-wide radiation research.

Privately, the scientists are telling the AEC that the cancer danger could be far worse, without even considering the separate hazard of irreparable genetic damage to unborn generations. Their proposal: An immediate reduction of permissible maximum radiation to no more than one-tenth the present limit.

The proposal, first made to an engineering group in October 1969 and since presented to Congress and numerous others, has touched off a storm of debate. The AEC issued a rebuttal, arguing that false assumptions were used to arrive at key statistics. A group of 29 leading radiologists and other scientists, including some at the Lawrence Lab, denounced the Gofman-Tamplin findings in a letter to the Joint Congressional Committee on Atomic Energy. The group backed the present radiation standard and expressed concern over "the alarmist views of a tiny minority of experts."

Under pressure from a subcommittee chaired by Sen. Edmund Muskie (D.-Maine), the Government's Federal Radiation Council announced recently it will undertake its first major review of radiation standards in a decade. Health, Education and Welfare Secretary Robert Finch, who heads the council, says the Gofman-Tamplin figures "probably overestimate the risks" but new research findings warrant a broad review of present standards.

Meanwhile, ecologists and others are using the Gofman-Tamplin studies and personal appearances by them to lend authority to a widening citizens' battle against radioactive pollution. In many areas of the nation, the foes of radiation are scoring gains or at least challenging the AEC.

In Eugene, Ore., a citizen's group collected enough

signatures to force voter reconsideration of a $225 million bond issue for a city-owned nuclear power plant. The bond issue was passed in November 1968 "before the problems involved in nuclear power generating became widely known to the public," says Mrs. Aaron Novick, secretary of the protest group and wife of a University of Oregon molecular biologist who worked on the Manhattan Project. In the new vote on the power plant, an amendment was passed that will force at least a four-year delay in the project.

In Gilliam County, Ore., Dr. Tamplin appeared to testify against an expansion of radioactive waste storage by a private company, Chem-Nuclear Services. The Oregon Environment Council, a group that includes Sierra Club chapters, also opposed the expanded storage license and raised conflict-of-interest questions about a Chem-Nuclear regional manager, Dr. Gary Farmer. Dr. Farmer approved the license while director of the State Radiation Section in June 1969, then quit to join Chem-Nuclear in September. Dr. Farmer told the state Board of Health that his decision to approve the license came before he considered joining the company.

The state of Minnesota is ready to go to trial against the AEC and Northern States Power Co. in a key lawsuit being watched by conservationists and regulators. The suit arose when the state, guided by advice from a former U.S. Public Health Service radiologist, set radiation emission limits on a nearly completed Northern States Power plant near Minneapolis that are only 2% of the AEC-allowed level. The AEC argues that Uncle Sam has preempted this regulatory field, while Minnesota is asserting that any Federal safety standard should be a guide that a state may toughen if it wishes. More than a dozen states have joined Minnesota to oppose the AEC.

Colorado is becoming a major battleground for opponents and proponents of nuclear energy. The opposition is led by the Colorado Committee for Environmental Information, a group of scientists from the Boulder area. The group is headed by H. Peter Metzger, an energetic, 39-year-old biochemist who works for Ball Brothers Research Corp., an aerospace contractor. He calls radiation "probably the most crucial pollution problem because the damage is irreversible," noting that plutonium, an ingredient of nuclear weapons, has a half-life of 24,000 years. This means half of any plutonium that pollutes the environment now will still be around in 24,000 years.

The accidental release of plutonium at the AEC's nuclear weapons plant at Rocky Flats near Denver brought a major confrontation between the environmental council and the AEC. The council's scientists studied soil samples for several miles around the plant following a $45 million fire there in May 1969. They found sites where "between 100 and 1,000 times" more plutonium was measured "than there would be if good containment practices were continually maintained." The plutonium levels found might or might not constitute a hazard to people, they added.

Officials of Rocky Flats, a 3,600-worker plant run by Dow Chemical Co., didn't collect soil samples until the council's report stirred wide criticism. Now they run regular tests within a 10-mile radius, and a spokesman says the plutonium escaping into nearby soil is only 10 times—rather than 100 or 1,000 times—the level one would expect in soil that had merely natural background radioactivity. The excess probably came from leaky storage drums at the plant, he adds. The spokesman, Dr. James Willging, says this level is "no more than one-fiftieth of the alarm-level" and is "totally harmless to the best of our knowledge."

Dr. Gofman becomes highly upset at such statements, and a key part of his attack on the AEC has been aimed at statements by many of its officials that small amounts of radiation are safe. "It's an outrageous lie that only high doses of radiation are harmful," he says. "At least 95% of the scientific community say all radiation is harmful or at least say the case is unproven."

Another bitter public fight has erupted over the use of uranium mine "tailings" or wastes in western Colorado. The tailings were purchased by contractors to use as earth fill beneath some 3,000 homes and other structures in communities such as Grand Junction and Uravan. Health authorities later discovered that the tailings have raised radioactivity levels in some of the homes to many times the permissible level. Dr. Metzger estimates that almost half of 120 homes identified as radioactive in Grand Junction exceed the allowed maximum level.

The uranium tailings contain radium, which gives off a gas called radon. Decay products of radon gas can be highly dangerous and are linked to a sharp increase in lung cancer among Rocky Mountain uranium miners. The miners now work under sharply reduced maximum exposure standards. Meanwhile, state and Federal officials are pondering what to do about all the houses built on mine tailings. So far, only two families have been forced to vacate their homes.

No recent nuclear issue has caused more controversy than the Plowshare program. Although there have been only about two dozen experimental Plowshare blasts thus far, some industry sources foresee thousands of underground nuclear detonations in the next few decades if Plowshare is to fulfill its promise of tapping mineral riches.

The American Civil Liberties Union and a Colorado

scientists' group challenged the AEC in a legal fight recently over Project Rulison, a 40-kiloton Plowshare experiment in western Colorado to stimulate natural-gas production. Government experts have estimated that such nuclear "fracturing" could free enough gas trapped in imporous rock to nearly double the nation's gas supply.

Rulison's opponents persuaded a Federal court in Denver to overturn the AEC's claim of immunity from citizen suit. Federal Judge Alfred Arras allowed the ACLU and others to sue for an injunction to prevent reopening the gas well but then denied the injunction itself. However, the judge retained jurisdiction of further safety disputes and ordered the AEC to make periodic public reports on test data from the well. He explicitly rejected arguments by Dr. Gofman and Dr. Tamplin and upheld present AEC radiation standards.

Environmentalists claimed at least a limited victory in the suit. "The suit means everyone attempting future Plowshare explosions may face law suits," says Dr. Metzger. That view is shared by at least one Plowshare proponent, Donald Gilbert, director of Arizona's Atomic Energy Commission. "I'm sure the (legal) questions are going to make a lot of companies hide and watch for awhile," he says.

There seems little doubt that the Plowshare program, already troubled by a budget cut to $8 million in the 1971 fiscal year from $14.5 million in 1970, now faces uncertainties. Kennecott Copper Co. at one time talked enthusiastically about Project Sloop, an experiment in blasting loose low-grade copper ore with nuclear explosives through which acid would be poured to leech out copper. The experiment, it said, could lead to nearly a doubling of copper supplies. But now plans for Sloop are "in mothballs" due to Federal budget cuts, a Kennecott spokesman says. "We still consider the project has merits," he adds.

Columbia Gas System Inc. already has felt the sting of public reaction to nuclear projects. It canceled its Plowshare program to explode a nuclear device near Renovo, Pa., to carve out a huge underground reservoir for gas storage. The move came after the Renovo Daily Record mobilized opposition and a retired forester in the area led an anti-Plowshare campaign that gathered 25,000 signatures against the project. Dr. E. J. Sternglass, director of radiology at the University of Pittsburgh, "gave us a lot of support to fight it from the radiation angle," says Bennett Shaffer, editor of the Renovo paper.

Dr. Sternglass, along with the Lawrence Lab scientists, has raised the strongest alarms against radiation hazards. He calculated that world-wide nuclear test fallout has caused 400,000 prenatal or infant deaths. The Sternglass report shook the AEC and was received skeptically by many independent scientists. In early 1969, the AEC asked Dr. Tamplin and his associates to prepare a rebuttal. This was before Dr. Tamplin broke openly with the commission. The biophysicist studied the Sternglass data and concluded it contained a statistical error; only 4,000 deaths, not 400,000, could be attributed to test fallout.

From there the story gets fuzzy. Dr. Tamplin claims the AEC ordered him not to publish the 4,000-death figure because it didn't want the public to get the idea anyone is dying from AEC activities. An AEC spokesman denies this. In any case, the Tamplin report was published, and he and Dr. Gofman soon broke entirely with the AEC hierarchy.

The two Lawrence scientists until recently enjoyed a high status in the Lawrence Laboratory in Livermore.

Dr. Gofman became medical director of the lab in 1953 and 10 years later was named to two key posts: Associate director of the entire lab and director of its new biomedical division. For the latter, he hired Dr. Tamplin as an associate.

Dr. Gofman resigned both his administrative posts at the lab before the current controversy broke out. He says he wanted to devote his time entirely to research. Dr. Tamplin remains a group leader in the biomedical division but asserts that all his scientific help has been transferred to other divisions and that some travel expenses, such as a recent trip to Washington to defend his research, are no longer paid by the AEC. A lab spokesman replies that Dr. Tamplin voluntarily gave up his staff and that the AEC can't finance travel for "propaganda" purposes.

Drs. Gofman and Tamplin began their public fight by estimating 16,000 cancer deaths might be caused if the entire U.S. population received the maximum permissible dose of 170 millirads yearly from man-made sources, excluding medical uses. (A millirad is a standard measure of radiation; most medical damage in humans is observed at a level several hundred times that amount.) Then they doubled the death estimate to 32,000 in Congressional testimony. In a memo marked "highly confidential" to Dr. Michael May, director of the Lawrence Lab, Dr. Gofman in December 1969 said he would guess the most pessimistic but still possible assumptions about radiation suggest that "the expected annual excess deaths from Federal Radiation Council guideline exposure will be closer to 250,000 deaths annually than to 16,000."

Dr. May calls the 250,000-death estimate "very extreme and kind of ridiculous." One critic of the Gofman-Tamplin studies, Dr. John Storer, scientific director for pathology and immunology of the AEC's Oak Ridge National Lab, told the Joint Atomic Energy Committee that various erroneous assumptions made by the Lawrence scientists resulted in their over-estimating cancer risks "by at least a factor of 100." This means 160 to 320 deaths rather than 16,000 to 32,000 might be caused by radiation at the maximum permissible dose.

The scientific evidence that underlies all this argument is complex and, most scientists agree, far from complete. Drs. Gofman and Tamplin assert there is no threshold, or limit, below which radiation is harmless. They assume human cells can't repair themselves after small or slowly administered doses of radiation; they assert that all forms of cancer likely can be caused by radiation; and they claim a "doubling-dose effect" from radiation, meaning that a given amount of radiation that doubles the normal or "spontaneous" incidence of one kind of cancer will double the incidence of all types of cancer.

Dr. William Bibb, a member of the AEC's division of biology and medicine, is among scientists who reject all these assumptions. "We cannot and I'm not sure we'll ever be able to demonstrate a harmful effect at very low doses," he argues. He cites studies on rats that he says show longevity improved with very small radiation doses, for example.

The scientific debate is far from ended. An article in Science, the magazine on the prestigious American Association for the Advancement of Science, conceded that the Gofman-Tamplin estimates might be right. "The inability to readily disprove their work on scientific grounds dramatizes the tenuous role that science plays in the determination of radiation risk," it said.

Almost lost in the heat of debate is the fact that,

nationwide, present actual radiation falls well below the maximum permissible limit. One AEC scientist estimates Americans now receive less than 1% of the 170 millirad yearly limit from man-made, nonmedical sources. An AEC spokesman declines to predict when the limit might be reached, noting the uncertainties of the Plowshare program among other things.

Meantime, the AEC can be sure that the two dissident scientists here will continue to raise questions. Dr. Gofman ended his December letter to Dr. May with a postscript: "Lawrence Radiation Lab and the AEC together insisted I become a watchdog of the public interest," he wrote. "Some watchdogs bark."

—HERBERT G. LAWSON

1970

New York's Dirty Air

IN Central Park, on a knoll just behind the Metropolitan Museum of Art, stands an unintended monument to the effects of air pollution.

The monument is a 224-ton granite obelisk known as Cleopatra's Needle, carved in 1600 B.C. and presented to the city in 1882 by the khedive of Egypt. The obelisk's makers cut hieroglyphic characters into all four of its sides, and the ancient writing was still plainly visible when it was brought to the park.

Today, however, there are markings only on two sides. Those on the south and west sides, which face prevailing winds and concentrations of air pollution, have been entirely obliterated. "Several inches of granite have been literally eaten off the obelisk by the chemicals in the air," says an official of the city parks department. "Ninety years in New York have done more damage to it than 3,500 years in Egypt."

Obviously, if New York's polluted air can eat away rock, it can hurt people. Doctors know it contributes to respiratory ailments, and they suspect it may cause cancer and even brain damage.

What's particularly frightening about all this is that New York's air is not much different from the air of many other U.S. cities—just somewhat dirtier because the city is bigger. Even traditional pure-air havens like Arizona, Colorado and Vermont increasingly are finding their air fouled with the same corrosive pollutants that, in sufficient volume, can damage monuments and threaten human health.

But because of the concentration of population and pollution here, New York City provides a model of sorts for those seeking to identify pollutants, trace their sources and sort out their effects.

The most obvious pollutants—though generally among the least dangerous to health—are tiny particles of dust from the streets and soot from incinerators, residential furnaces and industrial smokestacks. Scientists say most of these particles are too large to work their way into lung tissues, so they are more an aesthetic nuisance than a health menace.

There are a number of smaller particles, however, that are less visible but more menacing to health. They include asbestos fibers, which have been linked to chronic lung disease and cancer, and lead, which may cause brain damage in children and injure the nervous systems of adults.

Lead comes chiefly from auto exhaust. Asbestos is released constantly into the air from disintegration of automobile brake linings and from the construction of the city's massive buildings, where asbestos is sprayed on as a sheathing for steel girders. Dr. Cuyler Hammond, a former director of the American Cancer Society, has found that smokers working where they are also exposed to asbestos dust have a rate of lung cancer eight times that of the general population.

Altogether, the U.S. Public Health Service says, New Yorkers were bombarded in 1969 by 70,000 tons of "particulate matter."

At the same time, however, the city's air was invaded by 400,000 tons of sulphur dioxide and 1.6 million tons of carbon monoxide—gaseous compounds that in Manhattan regularly exceed the levels scientists consider harmful to health.

Sulphur dioxide comes from the combustion of fossil fuel (coal and oil) for energy and heating. In a typical year New Yorkers burn 2.5 billion gallons of fuel oil and 700,000 tons of coal. These fuels contain varying amounts of sulphur that are converted, on combustion, into sulphur dioxide and sulphuric acid.

Beyond such effects as causing nylon fabrics to disintegrate mysteriously, these compounds can be breathed deep into the lungs, injuring sensitive tissues. Prolonged exposure to even low levels of sulphur dioxide has been linked to heart attacks, mutations and cancer.

"We know that sulphur dioxide causes certain chemical changes," says Robert Shapiro, associate professor of chemistry at New York University. "If these same changes occur in human sperm cells, they could lead to mutations that would pose great threats to future generations. And if these changes occur in other cells, such as in the lungs, then sulphur dioxide might even be a cause of cancer."

In Manhattan sulphur dioxide in the air is often as high as .12 parts per million (ppm)—three times the .04 ppm mark established by the Federal Government as an acceptable level. In the city as a whole sulphur dioxide concentrations average .06 ppm.

The main pollutant in New York—as in most other areas—is carbon monoxide, chiefly from motor vehicles. Carbon monoxide is dangerous because it hampers the delivery of oxygen to the body's tissues. "At high concentrations it kills quickly," says the Public Health Service. "At lower concentrations it brings on headaches and a slowing of physical and mental activity."

As might be expected, carbon monoxide is most harmful to people who are exposed to high levels of it for long periods. And many people in New York breathe high concentrations of carbon monoxide over the entire day. New York state guidelines warn that levels should not exceed 15 ppm more than 15% of the time during an eight-hour period. But after four months of measurements in midtown Manhattan it was determined that carbon monoxide levels remained above that level all day every day.

During daytime hours when traffic is heaviest, the carbon monoxide level in Manhattan often soars to between 25 and 30 ppm, having an impact on the lungs equivalent to that of two packs of cigarets a day. In some areas of the city, such as the Lincoln Tunnel and the approaches to the George Washington Bridge, the carbon monoxide level reaches an astronomical 100 ppm—nearly seven times the "safe" level.

According to medical studies, exposure to this much carbon monoxide, even for short periods, can cause headaches, nausea and dizziness. After 90 minutes of exposure to only 50 ppm, the ability to make certain visual discriminations and time judgments is impaired—indicating that high carbon monoxide levels on streets and highways may be a factor in traffic accidents. Even exposure to levels as low as 15 ppm may have an effect on mental and sensory responses, researchers say.

The celebrated surliness of some New York City taxi drivers and policemen may actually be a symptom of carbon monoxide exposure, according to some authorities. Cab drivers and traffic policemen must be in the streets constantly, often in areas where concentrations of the gas are highest. "Every time I work days I wind up with a headache," says one cab driver, Robert

Uzak. "So I've asked to be put on nights permanently. I've been held up twice while on night shifts, but I would rather risk getting shot or stabbed than dying slowly from all the poison in the air."

Dirty air has caused other New Yorkers to make adjustments in their living habits, too. Few Manhattan residents, for instance, risk keeping their apartment windows open long. "I once went away for a day without shutting a window in my kitchen, and when I came back the room looked like the inside of a coal bin," says Phillip Rosenberg, who lives on Manhattan's upper West Side.

Some people say the city's dirty air has contributed to decisions to move away from New York. Joseph Wicherski, a former public relations man for the Chase Manhattan Bank, says air pollution is a major reason he decided to give up his job and move to the West Coast. "I never had a cold in my life until I moved here," he says. "But in the past year I've missed six weeks of work because of colds. It's this damned air pollution. Sometimes I take a breath and I can feel the dirt going down my throat."

The pervasiveness of air pollution in Manhattan is apparent to anyone who spends time there, and the sources are not hard to find.

The single biggest industrial polluter of the city's air is Consolidated Edison Co., which supplies electricity throughout the city and in suburban Westchester County. Con Edison says it burns 1.18 billion gallons of fuel oil, 3.7 million tons of coal and 79.2 billion cubic feet of natural gas a year to produce electricity. As a result, its 11 fossil fuel plants throughout the city hurl into the atmosphere each year 156,000 tons of sulphur dioxide, 6,400 tons of fly ash and other particulates and 113,300 tons of nitrogen oxides.

(Con Edison says, however, that its emissions of sulphur dioxide have dropped 55% and its emissions of particles have dropped 53% since 1967 due to a switch to low-sulphur fuels and a $150 million investment in pollution-control equipment. Con Edison also says it plans to eliminate the burning of coal almost entirely by the end of 1972 and is negotiating with suppliers to switch from its present low-sulphur fuel oil, which contains 1% sulphur, to a "low-low" sulphur oil containing one-third of 1% sulphur.)

The air near Con Edison's plants is usually the most polluted in the city. A long walk near the East River and 14th Street, where the company has a plant, can leave the stroller's neck begrimed with a layer of soot thick enough to scrape off with a finger.

Another major industrial polluter is the Phelps Dodge Corp. copper refinery in Queens, one of the five boroughs that make up the city. In its manufacturing process the plant belches out 100 tons of solid copper, copper oxides, fly ash and other particulates each year, city officials say.

(A Phelps Dodge spokesman says, however, that "there is no particular pollution at all" from the refinery.)

Most New Yorkers needn't look beyond their own homes or offices for a source of pollution. Most of the aerial garbage that floats over the city every day actually comes from incinerators and heating units in apartment and office buildings, particularly older buildings that lack pollution control devices.

The city's Air Resources Administration cites one structure on Fifth Avenue, just north of Greenwich Village, as a typical example of how apartment buildings contribute to pollution. The building is not large by New York standards—16 stories, 78 apartments with

400 rooms. But the amount of smoke, soot and dirt it sprays into the air is sizable.

The oil burner in the building, which uses 235,000 gallons of oil a year, pours 18 tons of sulphur dioxide and 140 pounds of particulates into the atmosphere annually, the Air Resources Administration says. The incinerator, in which 575 pounds of refuse are burned daily, produces 200 pounds of sulphur dioxide and 400 pounds of dirt particles a year. Most of the particles are too heavy to be carried away by the wind, and they fall within the immediate area of the building.

New Yorkers are apt to suffer most from air pollution during the months from October to February, when freakish temperature inversions are most likely to occur. A temperature inversion takes place when a layer of cool air is held under a layer of unseasonably warm air, trapping pollutants over the city and sometimes allowing them to build up to lethal proportions.

New York's worst temperature inversions in recent years occurred in late 1953, early 1963 and over the Thanksgiving weekend in 1966. Dr. Leonard Greenberg, former air pollution control commissioner for the city and now a professor at Albert Einstein College of Medicine, has studied all three incidents for the number of deaths recorded compared with normal periods. He has concluded that air pollution caused the deaths of 220 persons during the first inversion, 300 to 350 during the second and 168 during the third. (The worst disaster attributed to air pollution occurred in London in 1952, when some 4,000 "excess" deaths were recorded during a three-day temperature inversion in early December.)

City officials say that smog disasters like those of 1953, 1963 and 1966 will not happen again. Under an emergency plan adopted in 1967, the mayor has the power to restrict the operation of incinerators, the use of electric power, business operations and even city traffic to reduce pollution during a prolonged temperature inversion. A 38-station air monitoring network and a pollution alert system have been set up to warn of impending danger.

The warning system is part of an extensive program to curb air pollution launched in 1966. City laws now require the use of low-sulphur fuel for heat and power and require that all incinerators used in the city be fitted with equipment to curb the fumes and fly ash that otherwise spew into the air. The use of low-sulphur coal and oil has reduced the amount of sulphur dioxide in the air by more than 50% from the pre-crackdown era, though the level is still higher than the .04 ppm maximum recommended by the Federal Government.

But the installation of control devices on incinerators has been slowed by court challenges brought by real estate owners who contend the antipollution law has an unnecessarily hasty timetable. "Real estate people in this city want clean air as much as anybody," says Harold J. Traynor, lawyer for the Real Estate Board of New York, "but they want sufficient time to upgrade their equipment." It costs about $10,000 to install pollution-control gear on an incinerator, and building owners complain they need time to make such an outlay.

Meantime, of course, the city's air continues to be polluted by automobiles and by various other sources such as the city government's own refuse incinerators, and there are widely divergent views on how bleak the outlook is.

"The city has scored very substantial gains against air pollution," says Norman Cousins, the editor of the

Saturday Review and Mayor John Lindsay's adviser on pollution control. "There is no doubt that New York's air is much more breathable now."

But many New Yorkers share the view of Mrs. Linda Fosburg, executive director of a private group known as Citizens for Clean Air, who says that air pollution in the city "is worse now than it ever was." In a sense, both views are right, says Kenneth L. Johnson, New York regional director of the Federal Air Pollution Control Administration. While sulphur dioxide and particulates have been reduced, he says, the amounts of carbon monoxide and lead are actually increasing.

—Nicholas Gage

1970

The Disappearing Wolf

THE wolf, the big, bad animal that tried to do in the three little pigs and Little Red Riding Hood, is about to get done in himself. "Unless they're protected, it's entirely possible there won't be a wolf left in this country in five years except in Alaska and in zoos," says an Interior Department official.

That may sound good, but it's bad. For the wolf, say naturalists, is really a fine animal who seldom if ever has attacked human beings. Instead, he weeds out weak and diseased animals of the forest. They say he is vital to nature's scheme of things and is feared only because he is misunderstood. They say he needs a good public relations man.

"No other animal in history has had such a completely negative image in the human mind," says Douglas H. Pimlott, a professor at the University of Toronto and an authority on wolves. He says his children have played with wolves.

Wolves once stalked almost all of North America, but they have been killed off gradually by hunters seeking bounty and by farmers protecting their chickens and sheep, which wolves do eat. Now there are wolves in only half a dozen states in the U.S. Alaska still has thousands of the animals, but the other states probably have fewer than a thousand all told.

Wildlife officials are concerned. Minnesota, which has about half of all wolves left in the contiguous states, lifted its bounty on wolves in mid-1965, leaving Alaska the only state with a price on wolves' heads. Michigan, which only has about 50 wolves left, made it illegal to shoot wolves there in mid-1965, with fines of up to $100 and jail sentences of 90 days for violators. Wisconsin, which has fewer than 25 wolves, fines a wolf killer up to $100 and sends him to jail for up to six months.

There once were 26 subspecies of wolves in the U.S., but now there are only two—the small, red wolf, now confined almost exclusively to remote sections of Louisiana, and the larger gray or timber wolf, a 175-pound critter found in the northern states, Canada and Alaska. Both subspecies have been added within recent years to the U.S. Fish and Wildlife Service's list of Rare and Endangered Species.

The list highlights the plight of the animals. More important, persons engaged in predator control can't kill animals on the list; instead, they must be captured and relocated.

One problem the red wolf has is that he looks a lot like his mean cousin, the coyote. Even naturalists can't always tell the difference. "Ninety-nine percent of the public can't tell the difference between a red wolf and a coyote, so it's pretty hard to convince them the wolf is disappearing and needs protection," says Joe L. Herring, chief of the Louisiana division of fish and game. He is trying to get Louisiana to prohibit the killing of wolves.

Naturalists talk fondly of the wolf. Some even deny that he howls, insisting instead that he sings. They concede he is carnivorous. "But what," asks a Minnesota wolf fan, "would be the image of man if he didn't have butchers to do his killing so the rest of us could righteously look the other way?" Others say the wolf serves as a "natural pruning knife."

Naturalists also extol the wolf for his great faithfulness. Though most people refuse to believe it, he is one of the few mammals who helps care for his pups, and he generally sticks to one mate all his life. They also say the wolf is very smart and very cunning.

But naturalists haven't always loved the wolf. Theodore Roosevelt, who was a leading conservationist, called the wolf "the beast of waste and desolation."

But that description rankles the naturalists of today. Mr. Pimlott, the Toronto professor, and Russel J. Rutter, another Toronto naturalist, have co-authored a book, The World of the Wolf. In it, they get quite sentimental about the animal. They write:

"We believe that the future of this noble animal will depend a good deal on whether its image in the public mind fits Roosevelt's description or whether it can at last be seen for what it is—an exceptionally interesting part of our wildlife heritage and an outstanding representative of that wilderness which we are belatedly trying to preserve."

Wolves are already extinct in much of the world. They haven't been seen in England since around 1500, in Scotland since the 1700s and in Ireland since 1766. There never have been wolves in Australia, and people aren't really sure if the Eurasian wolf is even a wolf at all. Some say it is a jackal. There still are plenty of wolves in much of Russia, however—so many, in fact, that Siberian officials are conducting a wolf-extermination campaign.

—MITCHELL GORDON

1968

Changes in the Weather

WHEN people in LaPorte, Ind., want to schedule a picnic, they might do well to first check the production plans of the big steel mills in Gary, 30 miles to the northwest. The reason: When steel output goes up in Gary, more rain comes down on LaPorte. The precipitation is triggered by smoke from the mills, weather experts say.

Though scientists have long suspected dirty air might change the weather, they devoted little research to the subject until recently. But now stepped-up research indicates pollution is becoming so bad that it's altering the weather over hundreds of thousands of square miles of the United States and possibly the global climate as well.

"We're putting astronomical quantities of materials into the atmosphere, and there's no question it's affecting the weather," says Charles L. Hosler, a meteorologist and dean of the College of Earth and Mineral Sciences at Pennsylvania State University. "I'm afraid the changes are already greater than most people suspect, and there may be a threshold beyond which small changes in the weather could bring about a major shift in the world's climate."

Increased rainfall is just one effect of pollution on the weather. In some cases, the contamination actually reduces precipitation. Mounting evidence indicates, too, that dirty air helps raise cities' temperatures by preventing the escape of heat generated in the cities, prolonging their frost-free season as much as 60 days. It screens off sunlight and produces fog, hail and thunderstorms. And those who contend pollution is felt on a global scale say it has weakened trade winds, increased cloud cover and, in contrast to its effect in urban areas, lowered the earth's temperature.

The processes by which pollution changes the weather aren't fully understood, but scientists think they know generally what's happening. Man-made pollutants that are spewed into the sky each year—an estimated 160 million tons from the U.S. and 800 million tons world-wide—have loaded the earth's air with dust.

The turbidity, or dustiness, of the atmosphere in places with supposedly "clean" country air, such as northern Arizona, Yellowstone National Park and the Adirondack Mountains, has increased tenfold in the past decade, recent research has shown. Over the Pacific Ocean, the dustiness of the air increased 30% in 10 years, and the dust fall in central Asia, as measured by Russian scientists, is 19 times as great as it was in 1930.

The dust particles are often so small—much less than 1/25,000th of an inch in diameter—that they're invisible to the naked eye. But they are highly effective cloud-forming agents. They strongly attract water vapor, which condenses and freezes on them, forming ice crystals. These, in turn, form clouds. If sufficient moisture is present, the cloud droplets grow and eventually fall as rain.

Precipitation generated by pollution in this fashion isn't a freak occurrence. Researchers think it can happen any place where the air is dirty, and they have found several cases to support their theory. For instance, Belleville, Ill., 10 miles southeast—and downwind—of St. Louis, receives about 7% more rain annually than areas upwind of St. Louis where the air is cleaner. The added moisture falls mainly on weekdays, when pollution from automobiles and industry in St. Louis is heaviest. Belleville even gets more rain than St. Louis itself because the prevailing winds tend to blow the pollution away from the city. In the past 18 years, Belleville has had rainfall of at least a quarter-inch on 83 more days than St. Louis. All but one of those rainy days were weekdays; the other was a Saturday.

LaPorte, though, is by far the most dramatic example discovered so far. Its situation was documented by Stanley Changnon, a meteorologist with the Illinois State Water Survey, a state agency engaged in studying the area's water resources. Chicago's automobiles and factories, as well as the Gary steel mills, he explains, throw huge quantities of two rain-producing ingredients into the air—water vapor and the dust particles on which it can condense and freeze. Flying through the orange plumes of smoke streaming from stacks of the mills, researchers have found billions of ice crystals being formed.

Wind sweeps the polluted air southeastward. Many communities along the route feel the effects, but LaPorte bears the brunt. Between 1946 and 1967 it averaged 47.1 inches of precipitation a year. This was 19 inches or 47% more rain than fell at stations upwind of Chicago and the mills. There's no indication the pattern is any different now.

Mr. Changnon considers it highly unlikely that the soaking the town gets is due to chance or to nearby Lake Michigan. For one thing, during the 1955-65 period LaPorte had 31% more moisture than nearby weather stations that also would feel any effect from the lake.

Two other facts are even more convincing. La-

Porte's rain pattern generally matches the number of days of haze and smoke in Chicago, a rough indication of the air pollution level. And the town's rainfall rises and falls in concert with the area's steel output. "Peaks in steel production occurred seven times between 1923 and 1962, and all of these were associated with highs in LaPorte's precipitation," Mr. Changnon says.

Sometimes pollution has just the opposite effect. Clouds can become so over-seeded that no rain falls. This occurs when pollution generates so many dust particles that none of them can attract enough moisture to grow to raindrop size. One of the best-documented examples of this is in the sugar-producing area of Queensland, Australia. During the cane-harvesting season, the cane leaf is burned off before cutting and harvesting, casting palls of thick smoke over wide areas. Downwind of these areas, rainfall is reduced up to 25% from levels in neighboring regions unaffected by the smoke.

Vincent J. Schaeffer, director of the Atmospheric Sciences Research Center at the State University of New York, believes the same process has modified the character of rain and snow over the Northeastern U.S. during the past five years. "Instead of the downpours we used to receive, we get fine, misty rains and snows in which the drops are so small they tend to drift down rather than fall," he says. "We had more than 20 of these this year. We used to see only two or three a year." Rains and snows of this type can drift great distances, changing a region's precipitation pattern, Mr. Schaeffer says.

Air pollution can produce dense fogs, too. Penn State's Dean Hosler has studied and photographed extensively a massive fog bank 1,000 feet high that forms regularly near a paper mill in Lock Haven, Pa., and spreads 23 miles down the Bald Eagle Valley. "The fog appears about a third of the time, especially in fall and winter," he says. "Normally there wouldn't be any there at all. It appears on days when little or no fog is observed elsewhere in the state. This type of thing is repeated all over the country hundreds of times in different locations."

Scientists believe violent weather occurs more often with polluted air than with clean air. Downtown St. Louis during one eight-year period averaged five more days of thunderstorms a year than a rural area 13 miles upwind with less pollution, an 11% difference. Chicago's Midway airport had 5% more thunderstorms than O'Hare Field, 16 miles northwest in an area of presumably cleaner air. And LaPorte in 14 years has had 130 days of hail, four times as many as surrounding weather stations. "We really don't know how or why pollution causes these changes," says Mr. Changnon.

Dirty air is responsible for more subtle weather changes as well. Acting as a blanket, it is one factor—along with the massive amounts of heat-absorbing concrete—that keeps cities warmer than suburbs by retaining the heat from industry, autos, furnaces and other sources. Average annual minimum temperatures in the center of Washington, D.C., for instance, run around 49 degrees, five degrees higher than outlying areas. Temperature differences are greatest on weekdays, when pollution—mainly from autos—is greatest. In New Haven, Conn., the minimum temperature on Sundays differs from the countryside by only 1.2 degrees. During weekdays the difference is twice that.

Variations of that magnitude have a great impact on the frost-free season. Downtown Chicago has a frost-free period of 197 days; in surrounding counties it's about 160 days. To find open countryside with a frost-free time as long as downtown Chicago's, you would have to go 400 miles south. Meteorologists say almost none of the temperature difference between Chicago and surrounding areas is attributable to the heat-retaining effects of Lake Michigan. The lake, they say, warms the entire region—city and rural areas alike.

A blanket of pollution also means less sunlight—a year-round average of 15% to 20% less than surrounding countryside in many cities, it's estimated. During the four months of winter, the center of London gets only about 96 hours of sunshine, compared with as much as 268 hours in the open countryside. Even at that, London gets about 50% more sunshine than it did prior to the enactment in 1956 of smoke control laws.

Not all weather-changing pollution originates with autos or factories. Scientists say jet airplane emissions high in the atmosphere are affecting the weather in many parts of the world. In some areas, for instance, pollution from heavy jet aircraft traffic has noticeably increased the cloudiness, they say. "In the alley between New York and Chicago, below which I happen to live in Pennsylvania, we get cirrus clouds 90% of the time now," says Dean Hosler. "Most of them wouldn't have been there naturally before. They are formed strictly by the thousands of jet aircraft flying overhead."

Reid A. Bryson, University of Wisconsin meteorology professor, says the same phenomenon is occurring in the air lanes between New York and London. Cloud cover in this region has increased by 10%, he estimates, and he adds that cirrus clouds could well attain 100% coverage with the advent of the supersonic jet transports.

"For every pound of fuel a jet burns it releases a pound and a quarter of water. At those heights—20,000 to 40,000 feet—that will saturate a very large volume of air because it's so cold," Prof. Bryson explains. "It's like exhaling on a cold morning. The jets also put out dust particles, so they're providing a couple of things needed for cloud formation. And there's one jet approximately every six minutes across the North Atlantic. That makes for a lot of clouds."

Along with the greater dustiness of the atmosphere, the clouds are contributing to another phenomenon—the world-wide cooling trend, Prof. Bryson and others say. There's little argument that since 1940 the average annual temperature of the world as a whole has dropped by one-third to one-half a degree—which doesn't seem like much until you consider that the latest ice age was brought about by a temperature drop of only four or five degrees. The weather scientists argue that the dirt and clouds have increased to a point where they are reflecting away enough sunlight to override other factors that would tend to raise the temperature of the earth. One such factor is the growing amount of carbon dioxide in the atmosphere, resulting from increased combustion of fossil fuels. Carbon dioxide acts as a one-way filter, permitting the sun's rays to pass through but deflecting the heat given off by the earth.

Prof. Bryson concedes that meteorologists don't yet have enough information to predict what will happen to the climate in the future, but he adds: "The only basis we do have is to look at the past to see what did happen. Looking at the climate of the past, it is clear that small changes in the past 10,000 years had very large ecological effects and they can happen bloody fast. The end of the ice age took less than a century—kapow! It's fast, and that worries me because we don't know but what in a few years we could have a significant change that would disrupt our entire climate. And that includes where we grow corn and wheat."

—RICHARD D. JAMES

1969

Halting Highways

THE Government's modest attempts to limit highways that tear up homes or mar scenery appeared destined for scuttling when long-time road champion John A. Volpe took over as President Nixon's Transportation Secretary.

But instead Mr. Volpe is emerging as an even stronger opponent of disruptive highway projects than his predecessor, Alan Boyd, who launched the Government's curb-the-freeway drive. In decisions affecting highways from Texas to New Hampshire, Mr. Volpe is stopping projects that conservationists say would damage parks or sites of historical interest.

In San Antonio, the Secretary has delighted conservationists but irked Republican Mayor Walter McAllister and Texas highway officials by halting a road due to cut through two parks and by suggesting an alternate route. Mr. McAllister insists "there's no justification" for this change of plans now, and the state highway men have offered instead to complete the original plan without any Federal money. But Secretary Volpe spurns that proposal. The unbuilt part of the road has come to be known as "the Volpe gap."

In New Hampshire, Mr. Volpe has blocked construction of a 15-mile segment of Interstate Highway 93 that would have cut through picturesque Franconia Notch. Conservationists, displaying "Save the Notch" bumper stickers, have fought the project for several years; among other things, they fear the construction would damage the already crumbling rock profile that forms the Old Man of the Mountain, a scenic high point of the Notch. The Granite State's GOP governor, Walter Peterson, is trying to get the Secretary to reverse his decision, but apparently in vain.

"Freeways that adversely affect our environment cannot be built," Mr. Volpe declares.

The Secretary is also insisting that any highway projects that would demolish houses be held up until decent replacement housing is assured. Now, he predicts, road planners will "think a great deal more seriously about going through areas that involve taking houses."

These moves are alarming road builders, who expected a clear path when Mr. Volpe joined the Nixon Cabinet, while winning praise from a number of freeway enemies.

Francis C. Turner, head of the Transportation Department's Federal Highway Administration and a career Federal highway builder, makes no bones about disagreeing with the scope of his boss' new policy. But he also recognizes that with Mr. Volpe squarely on the side of many freeway opponents, there is no turning back.

Anthony W. Smith, president of the National Parks Association, expresses pleasure at the Secretary's moves "in refusing to approve highway proposals that would adversely affect park, recreational and surrounding areas." The Rev. Theodore M. Hesburgh, president of Notre Dame University and head of the U.S. Commission on Civil Rights, whose investigations have shown urban highways as one cause of minority-group unrest, calls Mr. Volpe's housing-replacement policy "forward-looking and long overdue."

Mr. Volpe will probably go much further in the months ahead to halt or relocate freeways designed to push through communities, parks or historical sites. While his department would prefer that state highway planners settle disputes on their own, the Secretary sees perhaps a dozen more in which he'll probably have to intervene. Some of these could involve highways affecting the Tinicum Marshes and Wildlife Refuge in Philadelphia, Minnehaha Park in Minneapolis and historical sections of downtown Charleston, S.C.

"I'm saying (to highway planners) that either they come up with decisions on their own to save the environment and help people or we might well have to take the project involved off the freeway system," he says.

While the states actually design the highways in the interstate system, the Federal Government pays up to 90% of the cost. Hence it can call off disputed projects or press for changes.

What's more, Federal officials expect state highway planners to proceed more cautiously on future projects. Many roads just won't be attempted, they say, because of the costs and complexities raised by the new restrictions on routes.

"I wouldn't anticipate much pressure from states to build new highways in metropolitan areas any more," says an official of the Bureau of Public Roads, a Transportation Department unit.

Because of various controversies, Highway Administrator Turner predicts that some 150 miles of interstate highways, probably costing more than $2 billion, will not be finished by the current 1974 deadline—if they're built at all.

Partly because of these delays, the cost of the interstate highway program is climbing and the time set for finishing it keeps lengthening. Mr. Volpe is expected to ask Congress to extend until 1978 the mammoth highway trust fund and the taxes going into it to pay for the interstate roads; this arrangement is now scheduled to run out in 1972. Cost of the program is now estimated at about $63 billion, up from the $41 billion projected when it began 14 years ago.

Volpe intervention is not the only reason highway plans are being upset. Despite the pro-highway stands of Mayor McAllister and Gov. Peterson, many other state and local officials are insisting that urban freeways not be built as planned or not be built at all.

Gov. Francis W. Sargent of Massachusetts has declared a moratorium on most road building in the Boston area, pending a new study of highway needs and social and environmental consequences. He says a planned "inner belt" linking Cambridge and Boston with an existing highway network might not be necessary. Mr. Volpe, his predecessor, thinks the road is needed, but he isn't about to pressure Gov. Sargent to build it.

Responding to opposition from five Cleveland-area communities, Ohio Gov. James A. Rhodes has ordered that a segment of Interstate 290 not be built through their region and may have it deleted from the interstate system. Local opposition to a segment of Interstate 278 in Newark, N.J., forced cancelation of the stretch of road. Town officials in East Hartford, Conn., oppose an interstate segment there.

"The question in some urban areas now isn't what route but whether to build any route," says Gov. Daniel J. Evans of Washington.

These reappraisals reflect a wave of public reaction against urban freeways. The opposition has intensified in recent years, reflecting the fact that states tackled the easier rural segments of the interstate system first and then later began trying to push freeways through built-up city and suburban areas. Now white suburbanites are joining black ghetto dwellers to protest roads, saying they are ugly, use valuable land, tear up homes, stores and parks, split neighborhoods and add to pollution and noise.

In a number of places, though, the local citizenry is deeply split over highway decisions. In San Antonio, many husbands favor the planned route through the two parks; many of their wives, including members of the San Antonio Conservation Society, support the Volpe-proposed alternative, which would cut through a wealthy community but would take only about four homes there. While some influential Republicans favor the original route, others oppose it. After some of Sen. John Tower's aides pushed for that plan, they heard from the conservationists and backed off.

Indicative of the local complications that can arise is a dispute over one leg of the proposed Baltimore freeway system. Originally it was to cut through the middle-class Negro neighborhood of Rosemont. A design team then proposed going through a white cemetery instead; a predictable howl followed. A leading alternative now is a route through a different, largely black residential enclave, taking 67 homes, says a Maryland Roads Commission official. Still, says a planner, "There's tremendous community opposition."

In Washington, resistance to roads is evident at the Government's top levels. High Nixon Administration officials have begun attacking the highway trust fund, calling it an excessive commitment to roads that has prevented allocation of funds for critical mass transit needs. Top White House domestic affairs staffer John Ehrlichman repeatedly urges Mr. Volpe to make antihighway decisions whenever necessary, the Secretary says.

So far, his decisions have dealt mostly with threats to parks and scenic beauty. In 1969 Mr. Volpe vetoed an interstate riverfront roadway through New Orleans, siding with historic-preservation buffs who contended it would mar the city's famed French Quarter. Mr. Boyd had ordered design changes to minimize such effects but stopped short of canceling it.

The Secretary has also given opponents of continued freeway building in Washington partial backing in their protest over displacement of housing and damage to scenic and historical attractions; he has recommended that some segments be restudied or rerouted.

His mandate that no roads be built without firm assurance of replacement housing could mean that highway planners may throw in the towel on some projects, considering the cost of finding new homes. Some 50,000 persons a year are uprooted from their homes by new highways. Replacement-housing problems have snarled highway projects in Charleston, W.Va., and Philadelphia, among other places.

There is ample evidence that the rising concern over the social and environmental effects of highways is making road building considerably costlier. A dispute over the proposed Richmond Parkway on New York's Staten Island, which was to penetrate a scenic area flanked by parks, camps and a golf course, has brought a proposed rerouting that's expected to cost $42 million, against $30 million for the original route. Baltimore's freeways now are tagged at $800 million, against $300 million when they were to be shorter and when prices were lower.

—ALBERT R. KARR

1970

Saving the Alps

CAN the French Alps survive the skiing boom? Nature lovers are beginning to doubt it. They're fighting a seemingly losing battle against a growing number of high-rise resorts that are transforming once-quiet valleys and mountainsides into centers of hectic pleasure-seeking.

At what's now the resort center of Lac de Tignes, for instance, hardy cross-country skiers used to make igloos for overnight shelters. Now throngs of visitors sleep at the eight-story Shamrock Hotel, fortify themselves at the Crepes a Go-Go snack shop and dance away their evenings at Le Scotch discotheque.

These facilities are just a start, moreover. A high-rise resort planned for nearby Val Thorens will accommodate 40,000 visitors, making it one of the biggest in the world. For ski slopes, the developers propose to take over part of an adjoining national park—much to the horror of conservationists.

"What are we headed for—pizza parlors on the top of Mont Blanc?" asks Maurice Controu, an officer of the French Alpine Club. The famous mountain, highest in Western Europe, is only 20 miles from Val Thorens.

The French Alps are not the only area of the world facing massive development by ski resorts, of course. Skiing is booming in many nations, and the trend to big resorts is spreading. A group of American developers recently visited Lac de Tignes looking for ideas.

"Seeing this project really gives me more confidence about prospects for much bigger ski resorts in the U.S.," says one of the developers, Peter Seibert, president of Vail Associates. His company developed a 4,500-bed resort in Vail, Colo., and plans to double its size. "The quaint chalet is on the way out," Mr. Seibert says.

Like their French counterparts, U.S. conservationists are battling the trend. The Sierra Club, a big American conservation group, currently is trying to block a plan by Walt Disney Productions to build a large resort, including ski facilities, in the Mineral King area of Sequoia National Forest in California. The development has been stalled for the time being by a suit filed by the club against the U.S. Government, contending it has no legal right to permit the development in the national forest.

For various reasons, French conservationists are less than optimistic about their chances of blocking development of the Alps—though that hasn't discouraged them from vigorous protest against the Val Thorens project and, in particular, the proposed use of national park land.

The group planning to build the resort is known as Societe d'Etudes Financieres et de Construction (SEFCO). It contends that annexation of part of the national park is vital to its venture. "Without this land, we wouldn't be willing to go ahead with the project at all," says Pierre Schnebelen, SEFCO's chairman.

The 6,000 acres in question—5% of the park—contains a glacier that would allow summer skiing after snow elsewhere had melted. The growing trend in France toward continuing winter sports into mid-August enables resorts to extend their seasons and boost profits.

"Nature lovers all over France are just scandalized"

by the Val Thorens plan, says Mr. Controu of the French Alpine Club. Conservationists note that the park is a wildlife sanctuary that doesn't even allow motorized vehicles. Building ski lifts and inviting thousands of skiers could be disastrous to the wildlife, the nature lovers argue.

They also contend that ski resorts may sometimes contribute to avalanches. They assert that cutting trees to open up ski slopes clears the way for sliding snow as well as for skiers. Moreover, when resorts move in, farmers sell their land. This takes away grazing livestock that keep the grass short. Short grass tends to hold snow in place, while long grass bends and helps the snow slide, experts say.

Mr. Controu's French Alpine Club and such other groups as the French Mountain Federation, the French Ski Federation and the Touring Club of France have led the protest against the Val Thorens project. They have encouraged many petitions by individuals directly to President Pompidou, and they have won the support of several influential newspapers.

The protests have apparently prompted the government to delay approval of the project, and President Pompidou has promised to consider environmental objections in making his decision. But conservationists fear they'll lose anyway, one reason being that they have little support from others within the government. Premier Chaban-Delmas, for instance, is known to support the resort plan.

For the government, the problem is economic. The number of skiing Frenchmen has soared to an estimated two million from only 400,000 a decade ago. With insufficient skiing facilities at home to meet the growing demand, many skiers spend their winter holiday francs in Switzerland or Austria. This hurts the nation's balance of payments.

To keep Frenchmen schussing on native snow, the French government is working hard to encourage the domestic skiing industry. The government also aims to attract foreign skiers.

The developers of the Val Thorens project deny that they will wreck the environment. They argue that tall buildings minimize the amount of land used and say that zoning regulations preclude uncontrolled expansion.

But the critics counter that the imposing new resort developments are a far cry from the old Alpine farming villages of rough wood and local stone that blend into the natural beauty. The resorts' urban-style luxury prevents visitors from appreciating the area's rugged grandeur, the conservationists say.

Indeed, there are plenty of man-made diversions in the Alps these days. Posters in Lac de Tignes advertise the "Big Wheel," a hairy pop music group "exclusive from Manchester" that's appearing at the Club Camille. In nearby Val d'Isere, home town of ski champion Jean-Claude Killy, a shop offers $60 Pierre Cardin ski slacks. For those who want a quick taste of the wilderness without venturing far, the local movie house offers a Western film.

Ski resort operators say today's skiers demand the lively bustle of a large resort. Says Michel Auberty, a SEFCO official: "Skiers 20 years ago often wanted to get back to nature and the wilderness. The new ones want crowds."

—ROGER RICKLEFS

1970

An Antipollution Bonanza

SEWAGE pumps. Sludge digesters. Smoke scrubbers. Baghouse filters. Flocculating chemicals.

Those items don't sound very glamorous, but they're some of the staples of the newest glamor industry: Pollution control. Such products will be essential in the multibillion-dollar effort to clean up the environment over the next few years.

Makers of the gear stand to clean up, too. As a result, everyone and his brother are trying to get into the business. Witness:

—Aluminum Co. of America, drawing on know-how acquired in fighting its own pollution problems, has established a division to develop and market techniques for abating air, water and land pollution.

—Coca-Cola Co., eying the fast expanding cleanup market, has agreed to acquire Milwaukee-based Aqua-Chem Inc., a water pollution control company, for more than $150 million in stock.

—Litton Industries Inc., the big electronics concern and defense contractor, already is expanding its less-than-two-year-old Environmental Systems Division, which is enjoying brisk demand for an air pollution monitoring system controlled by computers.

—Monsanto Co., the chemical maker, has formed a new subsidiary, Monsanto-Enviro-Chem Systems Inc., to design and build systems to control air, water and land pollution by industry. In five to seven years, Monsanto predicts, the subsidiary will have sales of $150 million to $200 million a year.

Besides attracting such blue chip names, the pollution-control field is spawning an array of new companies. Some of the new concerns, like Envirotech Corp. of Palo Alto, Calif., are substantial companies that already have achieved a solid position in the industry. But stock analysts say some others are questionable ventures apparently set up to capitalize on what's becoming a hot stock-market trend. Besides a rash of new concerns with bewilderingly similar names based on the words environment, ecology and pollution, some older companies with only slim ties to the field are being promoted as pollution-control concerns.

"I've seen companies call themselves air-pollution-control companies because they make things like furnace filters," complains James G. Wilcock, president of Joy Manufacturing Co., a major producer of equipment to reduce air pollution.

The reason, of course, is the hope that a pollution-control label will help a company's stock rise in price.

But stock analysts think that the companies that are trying to attach themselves to this trend in name only won't last long. "In the water pollution field, you can call some company Ecological Garbage Corp., but it still is a sewer-pipe company," says Dorothy Fels, an analyst who specializes in pollution-control stocks for the Pittsburgh brokerage house of Parker-Hunter. She predicts an eventual "shake-out" in the industry.

Men familiar with the equipment needs of government and industry assert that companies that will benefit from the massive effort to clean up the environment will include makers of such ordinary items as pumps and pipe, along with construction contractors and cement producers.

But the makers of specialized gear obviously will benefit, too. An official of Wheelabrator Corp., a Mishawaka, Ind., company that makes all kinds of air-cleaning devices, predicts "conservatively" that industry sales of such items will rise to $344 million in 1973 from $165 million last year.

An official of Pittsburgh's Koppers Co., which, among other things, makes devices that give dust particles in factory exhausts a negative electric charge that traps them on positively charged plates, says its sales of pollution-control equipment have more than doubled in the past five years. "With all the attention being given to pollution," says Donald DeVries, a vice president, "this industry is going to grow at a rapid rate."

Just how rapid is unclear. President Nixon has called on state, local and Federal governments to spend $10 billion to construct sewage treatment plants. One study by the Federal Water Pollution Control Administration estimates $26 billion to $29 billion would have to be spent by 1973 to bring streams and lakes up to the clean-water standards set by the Federal Government. Nobody knows how much money must be spent to clear the air.

All this spending won't mean huge profits for the equipment makers, they caution. "The profits are overrated," says Mr. DeVries of Koppers. "Everyone thinks this is the greatest thing since 'the pill' and that they ought to get in on it," he says. As a result, he says, "every time you turn around there's someone new" in competition with you.

Similarly, L. L. Bott, president of Chicago's Nalco Chemical Co., says, "Everybody else has discovered the business we've been in for 45 years." Among Nalco's products are chemicals to purify water. (Ironically, Nalco got into the pollution-fighting business in the

'20s when it developed a treatment for water coming *into* factories from polluted rivers.) Nalco's sales have been growing at an annual rate of 22% for the past five years, and Mr. Bott expects this rate to accelerate.

Another fast-growing business is the production of instruments to measure how good—or bad—a job is being done by various devices to control air and water pollution. As laws setting specific, measurable limits on effluents are written and tightened, the need to monitor what's going up in smoke or down the drain is increasing rapidly.

Litton's Environmental Systems Division, for instance, is cashing in on the Federal Air Quality Act requirement that each of 57 "air quality regions" around the country must set specific clean-air standards and enforce them. In most areas, the standards-setting is still in progress, and enforcement will begin within a year or two.

The Litton division has sold more than a dozen complex air monitoring systems, which cost $1 million or more each, says Leland S. Webster, a Litton vice president. The system permits measurement of up to 16 factors (such as temperature, wind speed and carbon monoxide level) at 32 stations in a city or over an entire region. The data from the remote centers are relayed to a computer, where the information is evaluated, recorded and displayed, providing steady surveillance of areawide air pollution.

While Litton seeks ways to measure pollution, other companies hunt new ways to fight it. A subsidiary of Rollins International Inc. is building disposal plants for industrial waste in various cities and contracting with companies to haul their wastes to these plants. The plants convert the noxious gunk into an inert solid that can be used for land-fill, says an executive of Rollins, whose main business is leasing trucks and cars. The company plans to build 100 such plants in the next five years.

The Rollins plants will handle concentrated wastes, such as toxic sludges and acidic liquids that result from the customer's preliminary treatment of waste water in its own factory. These are especially hard-to-handle pollutants that pose difficult disposal problems. As evidence of the need for such a service, a Rollins official says a major manufacturer is trucking a hard-to-dispose-of waste from Minneapolis all the way to a Rollins plant in New Jersey.

Many other pollution-control ventures are in various stages of planning, and the technology is developing quickly. "The art is emerging very, very rapidly," says John C. Cleaver, chairman of Aqua-Chem, which is turning increasing attention to the business of fighting pollution.

One area that several companies are investigating is the development of pollution-abatement systems that will help pay for themselves by recovering useful by-products. For example, Monsanto, Combustion Engineering Inc. and Ionics Inc., among others, are seeking an economical means of removing sulphur dioxide from plant exhausts and converting that noxious gas into salable sulphuric acid.

Some systems to recover by-products already are in use. Zurn Industries says a system it installed at Owens-Illinois Inc.'s paper mill in Valdosta, Ga., is recovering nearly eight tons of lime dust from plant emissions every day. The dust formerly settled on the surrounding countryside. Now it is recycled into the pulp-making process. The value of the recovered lime is enough to pay for the $100,000 recovery system in a little over a year, Zurn officials say.

Wean Industries, a subsidiary of Wean United Inc., last fall formed an environmental control division to market acid-recovery and other antipollution systems to such industrial customers as steel mills. A sulphuric acid recovery unit being installed by Wean at Republic Wire Corp., Carteret, N.J., captures reusable acid and commercially salable ferrous sulfate from waste "pickle liquor," a corrosive pollutant left after steel is given an acid bath to remove scale.

"We expect the unit to pay for itself in about two years, and after that it will be all gravy," comments B. J. Mondi, president of Republic Wire. The company used to dump the pollutant into a stream feeding the Raritan River, but pressure from local authorities in recent years forced the company to haul away the pickle liquor by tank car; that cost up to $50,000 a year—money that will be saved now, Mr. Mondi says.

Occasionally, the pollution by-products find their way into some startling uses—such as baby food. Ionics says it has installed at several U.S. dairy plants units to remove ash from whey, a milky by-product of cheese-making that often is dumped into rivers, where it can cause pollution problems. Ionics says the de-ashed whey, when mixed with skim milk, has a composition remarkably similar to human milk and is being sold nationally as a formula for infants.

—JAMES P. GANNON

1970

The Internal Combustion Engine

THE internal combustion engine has been one of the greatest commercial successes of the twentieth century. It has nurtured two giant industries—oil and autos—spawned suburbs, generated superhighways, created employment for millions.

Nevertheless, the internal combustion engine could be on its way out. At the very least, Detroit's auto makers are going to make fundamental changes in it by the end of this decade—changes that almost certainly will make autos more expensive, while reducing their power and performance.

Public alarm over air pollution is the reason. After nearly two decades of fighting a rearguard action against stringent pollution control—arguing that no one had demonstrated the need for a big cleanup, then arguing that they couldn't afford such a cleanup and that it would be too costly—auto makers are bowing to the inevitable. Under increasing pressure from Washington and the public, Detroit is moving toward the kind of pollution control that seemed unimaginable a few years ago.

The cleanup is moving ahead on two fronts:

—A crash program to meet guidelines proposed by the Nixon Administration, which call for cutting emissions of pollutants by autos by up to 93% by 1980, regardless of the cost or effect on the performance of cars.

—A costly hedging operation that will give the auto makers alternate power sources should these efforts fail.

For the first time, industry leaders have vowed to switch to alternate power sources if necessary. Explains Henry Ford II, chairman of Ford Motor Co.: "We have a strong vested interest in the survival of the internal combustion engine, but we have a far stronger vested interest in the survival of our company."

Some critics of Detroit, including a group of Congressmen led by Rep. Leonard Farbstein (D., N.Y.), argue that nothing less than junking the internal combustion engine altogether will solve the problem. "The only way that we will be able to eliminate automobile air pollution in this country is by setting pollution standards not based upon what the inherently dirty internal combustion engine can achieve but on the basis of the cleanest propulsion systems, steam cars and gas turbines," says Mr. Farbstein. He and other Congressmen have recommended a complete ban on the internal combustion engine by 1978.

But Detroit has no plans to scrap its gasoline-burning engine without an enormous political and public relations fight. Its stake in it, in terms of plants, machinery, equipment, parts and service facilities, runs to billions of dollars. And Detroit is betting that its cleanup will succeed—and get the pollution fighters off its back.

But a successful cleanup would radically alter the engine in ways that insiders acknowledge might be very costly to both car owners and the oil and auto industries that serve them. Car owners will be hardest hit. Says Donald A. Jensen, Ford's top air pollution official:

"From here on out, the public is going to feel the effects of pollution control—both in their cars and on their pocketbooks."

Engineers estimate cars with cleaner engines are likely to have at least 10% less power. This means less zippy performance—and possibly trouble for Detroit efforts to sell option-laden high-powered "muscle" cars, already in trouble because of their high insurance costs.

Cleaner engines probably would consume at least 10% more gasoline per mile and require more frequent, more costly maintenance. (One research device being tested by a unit of Standard Oil (New Jersey) boosts fuel consumption 16%.) Especially on small cars—and on some bigger cars when the devices aren't working precisely right—some types of experimental pollution-control systems can cause a car to buck and surge. Moreover, such devices might add about 10%, or $200 to $300, to the cost of a new car, industry sources maintain. A Federal study, however, puts the increased cost of one probable device at only $116.

Despite its grumbling, Detroit has considerably toned down its earlier efforts to slow down the trend toward pollution control. Says Ford's Mr. Jensen: "We don't even try to argue the need anymore (for gradual, rather than once-and-for-all reduction of emissions). We're going to get it all out, whether it really has to be or not." Thus, in the laboratories and on its test tracks, Detroit is grappling with a whole array of pollution problems.

Of the five basic types of air pollutants listed by some researchers, the internal combustion engine spews out four. One source estimates autos are responsible for 8% of all particulates (bits of solid matter, including lead) in the air; 46% of the nitrogen oxides; 63% of the hydrocarbons, and 92% of the carbon monoxide. (The fifth air pollutant is sulphur dioxide, a deadly gas that comes largely from the burning of coal and some heavy fuel oils.)

Almost daily, scientists are adding to the list of problems caused by automobile pollutants. Recent studies indicate that constant exposure to low levels of carbon monoxide can increase mortality among heart attack victims and impair ability to perform mental tasks or make quick, accurate judgments—the kind necessary to avoid traffic accidents, for example.

New information about nitrogen oxides indicates they may be more dangerous to health than originally believed. Researchers at the University of California at Los Angeles have learned that nitrogen oxides reduce the oxygen-carrying capacity of laboratory animals' blood by up to 38%. Even relatively low concentrations of nitrogen oxides can initiate this process and also induce inflammation of the respiratory tract in laboratory animals. Higher doses can restrict breathing.

Nitrogen oxides, which often give the air a brownish tinge, act as fertilizers and thereby help induce a harmful overgrowth of algae in lakes and streams after they settle out of the atmosphere. "Each car is a little fertilizer plant," says ecologist Barry Commoner of Washington University in St. Louis.

As the list of problems grows, so do the doubts that the internal combustion engine can ever be cleaned up enough. Even with the stringent controls now contemplated, the increase in the number of cars on the road means that by the year 2000 the total amount of auto pollution will stop falling and start to rise again, according to one calculation. And at least some Detroit experts fear current efforts to control internal combustion engine pollution are approaching the limits of technology.

The Nixon Administration guidelines for 1980 represent "the end of the line for internal combustion engine control technology," says Charles Heinen, Chrysler Corp.'s chief emission control engineer. "If more than that is required, it's going to mean electric cars."

Existing Federal pollution standards deal with only two of the four auto pollutants: Hydrocarbons (unburned fuel that helps cause smog) and carbon monoxide. These standards have reduced hydrocarbon output in today's new cars by 80% from pre-1960 models and have cut carbon monoxide output by 60%. But pollution fighters say current standards don't go nearly far enough—and that they actually aggravate existing problems in some cases.

New York City offers a case in point. Austin Heller, the city's former air pollution control commissioner, says smog devices added to new car engines actually have increased the level of some pollutants there. He says these modifications were based on California-style freeway driving conditions. The devices cause cars to burn 10% more gasoline in New York's stop-and-go driving, thus at least partially offsetting the gains aimed for and actually increasing the output of the two pollutants not yet controlled.

The proposed Federal controls would go after all four auto pollutants and would require reductions that automotive engineers consider drastic. By 1980, through two-stage reductions, hydrocarbons would have to be reduced an additional 91%, carbon monoxide another 80%, nitrogen oxide 93%, and particulates (primarily lead compounds) 90%.

Taking the first major step toward meeting these guidelines, oil companies have pledged to remove lead from gasoline, and the auto makers promised to design most of their engines to run on lead-free gas starting in 1971.

Lead originally was added to gasoline to increase its octane, or burning efficiency. This allowed Detroit to design more powerful engines. But leaded gas forms toxic compounds that escape out the tailpipe, promotes the formation of hydrocarbons in the exhaust and interferes with advanced devices that could be used to control hydrocarbons, carbon monoxide and nitrogen oxide.

Oil men claim removing lead could boost the price of gasoline one to four cents a gallon and reduce gas mileage. But auto men counter removal will solve most of the particulate problem and slightly reduce hydrocarbon emissions, as well as extend the life of pollution-control systems. These systems would attack remaining auto pollutants.

Several approaches to pollution control are being tested. One is "exhaust recirculation"—rerouting a small part of the burned exhaust gases through a network of tubing back into the cylinders of the engine. Since the rerouted gases contain little combustible material and take up space that otherwise would be occupied by fresh fuel, the result is to lower combustion temperatures inside the engine. This, in turn, inhibits formation of nitrogen oxides.

Another approach would destroy excess hydrocarbons and carbon monoxide by burning them at 1,500 degrees in large tank-like devices slung on each side of the engine. This "afterburner" or "thermal reactor" approach also requires a football-sized pump to force fresh air into the chambers. The system would increase gasoline consumption and possibly cause cooling problems for the engine generally. Also, it would require special metals that could resist high temperatures and elaborate devices to guard against damage to the afterburner should too much fuel be fed into it.

A third approach would force exhaust gases through catalytic mufflers containing chemicals to neutralize the pollutants. In some experimental systems, an air pump is needed; one Ford system would use a small electronic computer to regulate the flow of gases. Among the technical problems this system raises: How to keep the catalytic material from crumbling under the heat and the jolts of normal driving?

Auto engineers say none of these devices, by itself, would cut pollution to the proposed Federal levels. But they hope that combinations of these systems and pollution-control devices already in use would do the job. Both Ford and General Motors say cars on their test tracks can meet the first stage of the new guidelines, which become effective with 1975 model cars.

Despite this optimism for the short term, the uncertainty about the long-term outlook for success in curbing internal combustion engine pollution persists. It helps explain the stepped-up research on alternative power plants.

Once Detroit scoffed at such potential substitutes as steam engines or cars powered by electricity or gas turbines. But today GM says it is spending 50% more on studies of such alternatives than on its work to clean up the internal combustion engine. Such research will get a further boost from a recently announced Federal program to encourage study of alternate power systems.

On paper, at least, the steam engine appears highly promising. A burning unit converts fluid in a boiler into vapor; the vapor pressure drives pistons and powers the car. Fuel isn't exploded as it is in the internal combustion engine; it is burned, and largely consumed, in the burner unit. Thus, the steam engine emits about 93% less hydrocarbons than today's internal combustion engine, 96% less carbon monoxide and 93% less nitrogen oxide. The system is quiet, simple and cheap to operate.

But auto men flunk the steam engine because of a number of special problems it creates. The engine itself is expensive, bulky and heavy, has major lubrication problems and requires auxiliary power to operate such accessories as the heater and air-conditioner when the car isn't actually moving. There also are problems in keeping the fluid from freezing and in finding a compact condenser to convert the steam back to fluid again fast enough to keep pace with the needs of the boiler. (Inventor William P. Lear recently abandoned attempts to develop a practical steam engine for autos, although he's continuing work on one for buses and trucks.)

Detroit also rules out all-electric cars, at least for the near term. One problem is that conventional lead-acid batteries wear out too quickly, and a practical substitute has yet to be developed. Another problem is that if everybody had electric cars, demand for electric power would surge (electricity would be needed to recharge the batteries). This, in turn, could cause a sharp increase in air pollution from the necessary oil and coal-burning utility plants. (Nuclear power stations would not pollute the air, but they do present an unknown radiation hazard and also could degrade lakes and streams and damage the aquatic life in them by releasing vast quantities of heated water.)

For the near future, auto men regard the gas turbine engine as the most practical alternative. The turbine burns fuel steadily, nearly completely and in a precisely controlled manner. It puts out about 88% less unburned hydrocarbons and 85% less carbon monoxide than the internal combustion engine. (But so far at least, experimental turbines put out more nitrogen oxides.)

Researchers say, however, that turbines consume more fuel, accelerate poorly and might cost three times as much as conventional engines because the exotic metals needed for some parts are in short supply. These problems contributed to Chrysler's decision several years ago to shelve its turbine car project, even though it had built 50 turbine cars for testing.

If the bugs in turbines can't be eliminated, Detroit might turn to a "hybrid" system—a combination of a small, relatively pollution-free internal combustion engine and a battery-powered, pollution-free system that could be used to operate the car in congested areas where pollution control is critical. The small engine would be used to recharge the batteries during highway driving. Top driving speeds possible under such a system would be about 60 miles per hour, and the range on battery power would be about 100 miles. Such a dual power plant would be more costly than the conventional engine, in part because it requires more controls.

—CHARLES B. CAMP

1970

Reclaiming Refuse

SIX years ago Victor Brown came up with what he thought was a progressive—and profitable—idea. He would form a company to build processing plants capable of shredding and mechanically separating trash into its basic components of paper, metals, glass and other refuse, and then he would sell the recovered materials back to industry to be used again.

That way he would be making money from both the city whose garbage he handled and the companies that bought the separated trash. And he would also be performing a valuable service because the garbage would be kept out of incinerators and landfills, and resources would be saved through the reuse of the materials.

Today Mr. Brown is president of Metropolitan Waste Conversion Corp., which operates a plant that processes 25% of Houston's garbage. He charges Houston $4.11 a ton to handle the garbage, which is separated into paper, metal and a combination of crushed glass, yard refuse and food waste for garden compost.

But Mr. Brown is frustrated and disappointed, and he is beginning to feel he may be slightly ahead of his time. The reason: He's losing about $2 on each ton of garbage he handles because he can't sell most of the materials he salvages.

Of the 2,000 tons of garbage Mr. Brown handles each week, for instance, 1,200 tons consists of paper. But he can sell only 200 tons. "It's good solid paper—paper that's only been used once," says Mr. Brown. "It represents trees and a lot of other resources, and we're throwing it away and burning it."

Mr. Brown's business is known as "recycling." In recent months, with the surge of public concern over environmental issues, more and more government officials, business leaders and conservationists have pointed to recycling as a fundamental step toward alleviating such problems as pollution and the depletion of resources.

But, as Mr. Brown's experience indicates, several hurdles must be overcome before recycling is likely to become a routine, widely accepted process. At the moment, any broad move to recycling seems to be blocked by a complex set of factors, including unfavorable economics, technological shortcomings and restrictive government regulations.

Recycling is far from a new idea. Many metals and large quantities of textiles and rubber once were routinely collected by scrap dealers and reprocessed. But in recent years rising costs of collecting and processing used materials have discouraged their use.

About half the copper, lead and iron used in the U.S. is still recycled, but only about 30% of aluminum and 20% of zinc are reused. Less than 10% of textiles, rubber and glass is reprocessed nowadays. Of paper, the largest component of municipal waste, only about 20% winds up being used again.

The effects of recycling on conserving natural resources are particularly evident in the case of paper. The Association of Secondary Material Industries, a trade group, estimates it takes 17 trees on the average to produce a ton of paper. Of the 58.5 million tons of paper used in the U.S. last year, 11.5 million tons were recycled—meaning that 200 million trees did not have to be cut. But if 50% of the paper had been recycled, the association figures the cutting of another 300 million trees could have been avoided.

A number of projects and experiments have been launched recently to investigate possibilities for recycling. At the University of Missouri scientists are testing the feasibility of extracting glass from garbage and crushing it for use as an aggregate in asphalt paving. In San Francisco and in Madison, Wis., the public has been asked to separate its newspapers from other trash so that the papers can easily be collected and recycled. Officials in both cities say the public's cooperation has been greater than expected.

But advocates of recycling say far more work will have to be done before recycling begins to have any significant effect on environmental problems. "I think the approach up till now has been totally unimaginative," says Merril Eisenbud, professor of environmental medicine at New York University and former head of New York City's Environmental Resources Protection Administration. He advocates government subsidies to encourage industry to become more involved in recycling.

The Federal Government would become heavily involved in recycling activities if legislation now pending in the House and Senate gains approval. Bills in both houses provide for spending some $500 million in the next few years to support research and the building of recycling facilities by local and regional governments.

Recycling advocates aren't sure if the pending legislation is the real answer, however, since it places most of its emphasis on technology and tends to ignore economic factors. A closer look at Victor Brown's operation in Houston shows how technological and economic problems are intertwined.

In planning his Houston plant, which was built

more than three years ago, Mr. Brown figured most of his recycling income would be from the sale of scrap paper to paper companies. But so far he has been unable to sell any of his paper to paper companies; the 200 tons he sells each week go entirely to the construction industry to make building materials. "We have to fight just to maintain that small market," Mr. Brown says.

Mr. Brown contends the paper industry has rejected his paper because paper companies have such a heavy investment in woodlands and in pulp-making equipment that they simply aren't interested in recycling—an assessment that at least one paper industry executive concedes is partly true.

" A lot of the companies are oriented to the trees," says John Schmidt, assistant manager of manufacturing for St. Regis Paper Co. "If you have a lot of land with trees, you aren't inclined to abandon that."

But Mr. Schmidt says there are also other reasons for the difficulties Mr. Brown has had in selling his paper. St. Regis has considered buying wastepaper from Mr. Brown but so far has rejected it, arguing that the paper is mixed in quality, contaminated by other garbage and too expensive to transport from Houston to the company's recycling paper mills in the Midwest and North.

St. Regis officials argue that technology isn't yet sophisticated enough either to separate paper according to quality nor to remove the odor of garbage completely. "When we get to that point, Victor Brown might have a product," says Mr. Schmidt.

Mr. Brown's difficulties extend beyond paper. He says he's capable of turning out 60,000 tons or more of compost a year, but right now he can sell only 5,000 tons annually to agricultural markets. His only success has been in the sale of metals, mostly cans, to the copper industry, which uses them as catalysts in the production process. As a result of his losses, which he says have amounted to about $2 million over the past three years, Mr. Brown is cutting back on his research and development in an attempt to reduce costs.

Besides the economic and technological problems such as those plaguing Mr. Brown, there is the problem of regulatory restrictions. M. J. Mighdoll, executive vice president of the National Association of Secondary Material Industries, argues that many scrap metal, paper and textile dealers have been forced out of municipal centers to less-convenient locations on city outskirts because their businesses are considered "unsightly."

Mr. Mighdoll also contends that export limitations on materials such as copper and nickel, considered vital to national needs, have restricted markets and thus discouraged recycling efforts. He also cites a 10% depletion allowance that provides a tax break to growers of timber as a deterrent to the recycling of paper.

Recycling advocates maintain that many of these factors will have to change before industry will take more interest in recycling. Richard Vaughan, director of the Federal Bureau of Solid Waste Management, urges that the Government "provide the same kind of incentives for recycling" as have been provided for the exploitation of raw materials. He observes, for instance, that freight rates for iron ore and pulpwood currently are lower than those for scrap metal and scrap paper, a situation he argues could be changed by Government regulation.

Recycling might also be encouraged by adding extra charges on disposable consumer products, making reusable products more attractive and by somehow penalizing manufacturers who shun recycled raw materials when they're available. Such penalties might be imposed through special taxes, though conservationists haven't come up with any specific proposals yet. "These penalties would force the producer and consumer to look for alternatives," says Michael Brewer, vice president of Resources for the Future Inc., a nonprofit Washington-based research organization.

Many of those active in recycling argue that once the economic problems are overcome, the technological obstacles will easily fall. "All of the exciting things are in technology and all the answers are in economics," says Harold Gershowitz, executive director of the National Solid Wastes Management Association in Washington, a trade group that represents private handlers of solid waste.

Mr. Gershowitz argues, "You cannot separate the need for technology from the need for markets." He suggests that the Government begin creating markets for recycled products by confining its own purchases to recycled goods. The same argument is echoed by conservationists. "If the Government would say it would buy only recycled paper, recycling paper plants would spring up all over the country," maintained Jerome Goldstein, executive vice president of Rodale Press Inc. in Emmaus, Pa., which publishes several conservation magazines. Mr. Goldstein says that he has asked his paper suppliers to seek out only recycled paper for use in Rodale publications.

—DAVID GUMPERT

1970

Persistent Pesticides

THE Government's heralded campaign against the perils of pesticides has produced a series of tough-sounding prohibitions. But these edicts are proving less hard-hitting than they first promised to be.

In February of 1970 the Agriculture Department suspended the use of a highly toxic mercury fungicide on grain seed. Three Alamogordo, N.M., children had suffered serious brain damage after eating meat from a hog accidentally fed seed treated with the chemical. Yet seed processors, by department agreement, are continuing to use their remaining stocks of the fungicide.

Two months later the weed-killer 2,4,5-T, suspected as a possible cause of birth defects, was banned for use around the home and its recall from retailers' shelves was requested. Yet an estimated six million spray cans and other small containers of the compound remain in the hands of local merchants, and surveys show that sales to the public are continuing.

Late in 1969 the Government authorization of DDT products for house, garden and certain other uses was canceled with considerable White House fanfare. Yet three companies—Carolina Chemicals Inc., Diamond Shamrock Corp. and Lebanon Chemical Corp.—still have the right to go on marketing the chemical labeled for these very uses, while their protests await action by the Agriculture Department's tortoise-paced appeals machinery.

These delays dismay many conservationists and consumer spokesmen. Harrison Wellford, an associate of Ralph Nader, complains that letting sales of these supposedly prohibited DDT products continue "in effect leaves no ban at all." He notes that the department has set no date for a public hearing on the companies' appeals and the Pesticides Act provides no deadline. "Here is the loophole in the act and the department apparently is taking full advantage of it," Mr. Wellford charges.

Despite the criticism, Secretary Clifford Hardin's lieutenants insist they're taking a tough stand on pesticides. They cite the initial actions taken against suspect pesticides as evidence that the department, long criticized for subordinating health and environmental considerations to the interest of pesticide makers and users, is cracking down. Moreover, current Government reappraisals of pesticide dangers indicate this trend will broaden in months ahead, threatening a sizable cutback in pesticide producers' $2.5 billion annual sales.

"The general feeling around here used to be: Don't offend the pesticide industry," says one top official. "And when it came to human health, the department was willing to risk a degree of hazard. But now we're taking the position that there must be a big margin of safety for the public."

Matching such words with the fate of the mercury seed-treatment compound seems difficult, however. Initially, officials sent out a warning that the chemical was an "imminent hazard" to human beings and animals and that it was being recalled from seed processors. But when a major producer of the compound, Nor-Am Agricultural Products Inc. of Chicago, protested that this was too severe, officials abandoned the recall idea.

Apart from any department softness toward the industry, current reform efforts must also contend with a quagmire of technical problems and procedural delays.

Still to be agreed upon, for instance, is a method of disposing of pesticides once they've been banned—in particular, the approximately six million containers of 2,4,5-T. Companies have delayed recalling their products until safe handling procedures are decided. A state-Federal-industry conference on the subject has been looking into this.

"The USDA has a bear by the tail on recalls, no doubt about that," comments James Dewey, chief of the pesticides program at Cornell University's college of agriculture and head of an informal panel developing disposal guidelines for New York state authorities. Incineration is considered the safest method of pesticide disposal, but the Cornell professor has discovered that no Empire State community now has an incinerator able to achieve the extra-high temperatures required or the equipment needed to clean emerging gases. Most communities in other states are also believed to lack such facilities.

Another problem is the administrative jumble created by the mounting backlog of appeals filed by pesticide manufacturers. Nine pesticide-use issues now are awaiting judgment by Agriculture Department scientific panels (only two of the needed nine have been appointed to date). A company that loses at the scientific-panel stage still has the right to a public hearing, and 12 issues are lined up for hearings. The combined total of these pending cases—involving some 45 companies in all—has doubled since the start of this year. A court appeal following department hearings could delay a final decision for two or three years altogether.

While appeals are pending, manufacturers are free

to continue marketing products whose authorization has been formally canceled. And the DDT producers are not the only ones enjoying this freedom.

Four makers of lindane vaporizers, used for insect control and deemed unsafe around food preparation areas more than 14 months earlier, were still selling their products for use in restaurant kitchens while awaiting first step in their appeal, a scientific-panel opinion.

To plug some of the existing gaps in pesticide regulation, the Nixon Administration has planned to offer a number of amendments to strengthen present law. Samples: Pesticide inspectors would be empowered to enter a producer's plant and take samples for chemical analysis to assure that products are effective and don't contain harmful adulterants. In the case of a product suspected of being dangerous, the Agriculture Secretary would be authorized to halt all sales for 90 days while the department weighs the scientific evidence and decides what final steps to take.

But even if Congress approves such changes (and industry lobbyists already are trying to blunt the planned Administration thrust), questions about the resolve of pesticide regulators will remain, according to skeptics.

Consider the Alamogordo case. State and Federal investigators quickly discovered that a hog slaughtered by the father of the poisoned children to feed his family had been fattened on a diet that included sweepings of waste seed treated with methylmercuri, a preventive for root rot and other soil-borne diseases. They also learned that some three months before the children's illness 12 other pigs owned by the family had died after being stricken with blindness and paralysis.

Agriculture Department pesticide regulators say

the incident was their first clue that methylmercuri formulations caused brain damage. (There had been warnings of possible danger by Food and Drug Administration officials, though.) The department promptly classified such products as a health hazard and suspended their authorization for use in seed treatment.

Actually halting the use of methylmercuri formulations proved another matter.

At first, the department decided that the mixtures were sufficiently dangerous that the four companies marketing them should be requested to recall them from "all product locations." Nor-Am Agricultural Products, which makes nearly all of the compound used in seed treatment, immediately halted shipments from its Illinois plant to some 50 distributors. But it balked at calling in stocks already in the hands of distributors and seed processors, then at peak operation in preparation for spring planting demand.

In late February 1970, Nor-Am representatives met with Agriculture Department officials in Washington to argue that recall "was not economically or technically feasible," says Lionel Hart, Nor-Am vice president. To meet safety requirements for shipment of poisonous materials would entail costly repackaging at possibly hundreds of locations, the company contended. Once assembled, moreover, the stocks would present Nor-Am itself with the problem of how to get rid of them without causing dangerous air or water pollution.

The safest alternative, company executives maintained, would be to continue coating the compound on seed until field supplies were exhausted.

Nor-Am's argument proved convincing. An Agriculture Department memo commented that "the hazard to the public would probably be greater from the returning of existing stocks to the manufacturer than from (their use) in accordance with presently accepted directions."

But Nor-Am still wanted permission to sell stocks on hand in company warehouses. When the department refused, Nor-Am filed suit in Federal court in Chicago to overturn the suspension (alleging in part that the Government hadn't linked the Alamogordo poisonings with its products). In April the court enjoined the department's action, and Nor-Am resumed shipments until the Government obtained an appeals court stay of the injunction. The higher court hasn't yet handed down its final opinion.

Despite this legal clash with Nor-Am, the department has long been accused of willingness to accommodate the pesticide industry. The House Government Operations Committee in 1969 charged that the regulators "consistently failed to take action to remove potentially hazardous products from marketing channels;" failed to cancel authorizations of ineffective or potentially hazardous materials as quickly as they should have, and approved "numerous" pesticide products over the objections of the Health, Education and Welfare Department.

In partial defense, Agriculture officials say pesticide regulation has been hampered by lack of sufficient men and money. As recently as 1964, the pesticides agency had only 11 inspectors across the country to check on an estimated 45,000 registered products. Since then, the inspection force has more than tripled. But Gus Conroy, newly promoted to enforcement chief, says he's still a long way from his goal of sampling each registered product for chemical analysis at least once every five years.

Perhaps a more fundamental handicap is the persistent belief that the Agriculture Department's first duty is to agriculture. But department science director Ned Bayley, a Johnson Administration holdover, began to challenge this premise in a serious way following last year's House investigation. One consequence was a restatement of pesticide policy by Secretary Hardin. Henceforth, he said, the department would encourage "those means of effective pest control which provide the least potential hazard to man, his animals, wildlife and other components of the natural environment."

Mr. Hardin's most significant act to date may have been to formally delegate to the HEW Department the full responsibility for deciding pesticide health questions.

On this basis, the Agriculture Department moved in April of 1970 against all home, aquatic and food-crop uses of 2,4,5-T even though officials deemed the evidence supplied by HEW—that heavy injections of 2,4,5-T caused birth defects in mice—to be rather thin. In a letter to Mr. Hardin, then HEW Secretary, Robert Finch contended that "a prudent course of action" should reflect the possibility that the weed-killer "may present an imminent health hazard to women of child-bearing age."

But this move hasn't satisfied the more militant environmentalists. A group headed by Mr. Wellford is petitioning the department for a total ban on 2,4,5-T products. This would affect use of the herbicide for rural weed control. This is the principal application for the chemical, and it was not covered by the April action.

There's dissatisfaction, too, about the effectiveness of the department's earlier orders. At a Senate hearing, Mr. Wellford testified that a survey by the Nader organization of 15 hardware and garden stores in the Washington area turned up five still selling 2,4,5-T products, even though pesticide companies supposedly have sent out letters warning retailers to stop such sales. A check by Senate staffers of 10 such stores in the Baltimore area turned up seven still offering customers 2,4,5-T products.

—BURT SCHORR

1970

Ecology vs. the Economy

FROM the local chamber of commerce to the canyons of Wall Street, a deep faith in economic growth is a fundamental part of the American tradition. As it nourishes each American's hope for a better life, its value becomes a sacred assumption. To question it is to threaten the American dream itself.

Thus it was no accident that President Nixon, in a State of the Union Message suggesting a need to "re-direct" economic forces to find ways of cleaning up the environment, took care to reaffirm his belief in economic growth: "Continued vigorous economic growth," he said, "provides us with the means to enrich life itself and to enhance our planet as a place hospitable to man."

Against this background, one of the thoughts unearthed in the current round of public commentary on the environment problem takes on disturbing dimensions. It is the notion that "there is a fundamental contradiction between economic growth and the quality of life," to use a phrase the President sought to discourage.

Now there is unquestionably some truth to the idea, and as it has dawned on assorted conservationists, ecologists and others, it has produced a theory: Since the U.S.' supply of natural resources is finite, open ended growth of an economy with an increasingly voracious appetite for them is intolerable; if our land is to continue to support life, economic growth must be halted or even reversed; Americans must somehow create a "no-growth" economy.

In the current atmosphere, with environment hailed as "the issue of the Seventies," and ecologists elevated to the covers of national news magazines, the no-growth economy idea shows signs of becoming another impressionistic cause, with advocates long on insights damning the destructiveness of affluence and short on any constructive suggestions for change.

As it has with other issues, there is danger that this kind of thinking will obscure the more difficult truth: Practical ways to resolve the obvious conflict of economy and environment are far from clear for a free society; our hope for survival is shadowed by our ignorance.

In more detail, the popular argument against growth goes something like this:

Man got into trouble with the environment when he began to harness and use energy on a large scale from sources beyond his own muscles. Damage to irreplaceable resources results both from attempts to produce the energy and the uses to which it is put.

Processes of extracting and refining coal and oil may destroy land and pollute the air and water. Attempts to harness streams for electric power may flood useful land and create geological imbalances; nuclear power reactors may raise the temperature of nearby water and produce radioactive wastes difficult to render safe.

And the availability of power from such sources is the key to the industry and technology which combine to damage the environment in hopelessly complicated ways. Indeed, technological advance seems to compound man's harmful impact on the environment.

To run an air conditioner, observes Wayne H. Davis, a University of Kentucky biologist, "we will strip-mine a Kentucky hillside, push the dirt and slate down into a stream, and burn coal in a power generator, whose smokestack contributes to a plume of smoke massive enough to cause cloud seeding and premature precipitation from Gulf winds which should be irrigating the wheat farms of Minnesota."

Therefore, the argument goes, there is delusion in hopes that further technology can solve the environment problem; an economy based on expanding technology is itself the problem. The solution requires limiting it or reversing it, rather than expanding it further, no matter what the purpose.

It is suggested that economically "developed" societies which rely heavily on technology should look to simpler, more isolated societies which don't. People who live in relatively primitive Pacific islands, for example, find their lives circumscribed by basic limitations not felt by people transfixed by the seeming miracles of technology.

Jerry Mander, a West Coast advertising man who has involved himself in conservation causes, points out that in Micronesia, islanders "simply don't think about . . . the idea that everything is possible." Rather, he says, they have "had to gain a good feeling for pacing the breadfruit production and the coconut eating. In some of these places, the highest crime is cutting down a coconut tree without communal permission."

Since ultimately the earth's resources are no more infinite than the Pacific island's, the argument concludes, more developed societies must learn to live by similar limitations; in effect, to curtail the economic and technological growth which changed human life, for better or worse, from primitive to modern.

The argument, so far as it goes, is bound to have a certain appeal—many Americans probably are not as comfortable with their affluence as they thought they would be. Still, beyond the obvious problem of achieving a slowdown without unacceptably severe economic consequences, the no-growth theory leaves a huge question unanswered.

Is economic growth the cause of man's tendency to destroy the environment at an accelerating rate, or is it only the reflection of a far more basic human trait? Is the desire to exploit non-human energy only the result of shallow materialistic pursuit of an ever higher GNP, or does it have deeper roots in human psychology?

Here it is useful to recall one of the more disturbing observations of the Austrian physiologist Konrad Lorenz in his book "On Aggression." Like certain physical anomalies afflicting lower animals, he suggests, the modern human tendency to a high degree of organization and a hectic way of life, both often justified in terms of technical or economic accomplishment, is actually a kind of grotesque overdevelopment which violates nature's usual laws by continuing to grow even though it may be harmful to man.

As is the case with some other species with improbably or dangerously overdeveloped characteristics, he suggests, human overdevelopment occurs because man now expends much of his natural aggressiveness in competition with others of his own species, rather than in protecting himself against other animals and the elements. And the aggressiveness is not subject to easy repression.

When this dimension of the problem is pointed out to advocates of a no-growth economy, some respond by citing certain hippies who abandon urban life for simple communal existence in rural areas as proof that man can return to harmony with the environment. Mr. Mander talks of publicity campaigns to create "an island psychology in everyone on earth."

Against Mr. Lorenz' insight, these responses seem inadequate. The hectic pace of modern men, he writes, "is unnecessary, for they could easily agree to take things more easily; theoretically they could, but in practice it is . . . impossible." Indeed, he observes, economic and technical overdevelopment in modern society causes many to be afflicted with the so-called "managerial" diseases—heart attacks, high blood pressure, ulcers, neuroses—reason enough for men to have turned away from it, had they been able to do so, long before environment became an issue.

Mr. Lorenz' ideas are useful to put the economic growth question in a realistic context, but they probably are not reason enough to conclude that the environment-economy conflict is hopeless. Rather, by emphasizing the importance of deep-seated human factors, they suggest that trying to end man's tendency to economic growth is something like trying to end his tendency to racism. Indeed, the lessons of the civil rights movement may be useful in thinking about policy approaches to the environment-economy conflict.

As the history of race relations in the Sixties showed, it is possible to achieve some broad human change, perhaps far more than once seemed possible; but it is not a phenomenon which can be controlled very closely either by any private groups or by government. Furthermore, the threat of widespread disaster may in the end prove the only truly effective catalyst.

For policy, such a situation implies a need for a kind of humility before the unknowns, a willingness to proceed in short realistic steps, rather than reaching for huge goals which future events may suddenly prove to have been disastrously oversimplified. The fact that environmental damage seems more suicidal than racial tension may make such a course nerve-racking; it does not render it invalid.

Specifically, a meaningful first step in resolving the conflict between environment and economy might involve policies designed to make the economic system accountable for the damage it does to the environment, much as the civil rights laws of the early Sixties sought to recognize the black man's claim to full participation in the political system. Like the civil rights laws, attempts to make the economy more accountable to the environment would only provide the beginnings of a solution. Still, if this were understood from the outset (as it probably was not in the case of the civil rights laws) the logical next steps might become apparent in a relatively short time.

To effect such accountability, according to economists who have begun to think about the problem, the costs of avoiding environmental damage will have to be worked into the pricing system, an approach hinted at by Mr. Nixon in his State of the Union Message and embodied in a water clean-up bill sponsored by Sen. William Proxmire (D., Wis.).

There are probably many ways to do this; the most obvious is to establish as a tax base the volume of waste matter disposed of in the environment by individuals and industry. This would provide an incentive to recycle wastes which now pollute the air and water and add the cost of avoiding the pollution to the price the consumer pays for manufactured products. Ultimately, higher prices might begin to reduce demand for products whose manufacture damages the environment; in other words, the mass re-examination of values necessary to the beginnings of broad human change might begin to take hold.

Al Kneese, a Washington economist employed by Resources for the Future Inc., a research organization concerned with problems of natural resources, contends that such an approach is the logical place to begin.

Right now, he says, "our most limited resource is the assimilative capacity of the environment. Getting that under control might solve a lot of problems." He tends to a certain skepticism over some of the more extreme "visions of the apocalypse" currently in vogue among environmentalists, and he goes on to add a warning for those who would base today's policies on oversimplified assessments of the future:

"There are some hard problems" posed by the environment, he says. "And yet we are not sure how much these are problems. We just don't know. That in itself is pretty frightening. But the man who claims he knows is dead wrong."

—DAVID C. ANDERSON

1970

Cleaning Up in Court

Do you have a legal right to fresh air?

A growing number of conservation groups and governmental agencies contend that you do—and that you likewise are legally entitled to clean waterways, clear blue skies and surroundings where a reasonable number of scenic wonders and breeding grounds for wildlife are left undisturbed. They contend, moreover, that it doesn't take a special law to guarantee your right to these things—they're simply yours.

That may sound like a novel idea, but it's inspiring a broad legal attack in courts across the nation against all sorts of degradation of the environment. Scores of lawsuits have been filed recently by conservation organizations, individuals and public officials determined to establish the right of everybody to pollution-free surroundings. Many suits contend that ordinary legal concepts of nuisance and negligence should be expanded to protect the environmental rights of the citizen, while others argue that such rights are protected by the Constitution.

In Phoenix, for example, two university professors and their wives are seeking $2 billion in damages from six copper companies that operate nearly all the copper smelters in Arizona. The suit charges that smoke from the smelters not only is injurious to health but also restricts visibility and "damages the natural beauty of the environment." The plaintiffs are asking that the damages be divided among the 700,000 residents of the Salt River Valley, which encompasses Phoenix.

In Alexandria, Va., a citizens association and two conservation groups have filed suit to block an apartment development that would require filling about nine acres of the Hunting Creek Estuary, one of a number of estuaries that shelter wildlife along the Potomac River. The suit, brought on constitutional grounds, contends that the development would damage a valuable natural resource of the people.

The Justice Department has filed a civil suit against Florida Power & Light Co., charging that the utility is creating thermal pollution of Biscayne Bay with heated water discharged by two of its power plants. Contending that the thermal pollution is "rapidly ruining marine life," the suit asked a court order requiring the company to set up and adhere to a schedule for ending the pollution.

There's ample evidence that the suits filed so far are just a start. The battle is increasingly taking on the character of a well-organized campaign coordinated by various environmental groups.

The Conservation Foundation, a Washington-based research and information group, has announced plans to publish an environmental law "reporter" to bring details of significant pollution litigation to lawyers. The American Trial Lawyers Association has formed an environmental law committee, and it has held seminars on environmental suits for lawyers, scientists, law students and others.

A group known as the Natural Resources Defense Council Inc., modeled after the NAACP Legal Defense and Education Fund, plans to bring pollution suits and to help others do so just as the fund has done in civil rights cases. The new group's organizers are a number of New York attorneys.

The defendants in environmental suits are nearly always corporations, and the specter of an explosion of such litigation has confronted U.S. industry with a huge potential for damage liability. Corporate executives also are worried about the delay and disruption of planned expansions that are challenged by environmentalists.

The potential cost to industry in existing claims is already sizable. "The amount in litigation over the country is staggering," says Frank L. Seamans, a Pittsburgh lawyer who has defended such companies as U.S. Steel, Aluminum Co. of America, Sun Oil Corp. and Monsanto Chemical Co. against pollution damage suits.

One suit, filed in Federal court in New York on behalf of "all the people of the United States" including "generations yet unborn," demands a colossal $30 billion in damages from eight producers of DDT for punishment and restitution of damage allegedly done to the environment by the pesticide. The suit was filed by Carol A. Yannacone of Patchogue, N.Y., wife of a lawyer who has been active in environmental causes.

Another potentially huge claim has been filed by the state of California against the four major auto makers and the Automobile Manufacturers Association. Charging that the companies and the trade association conspired to halt development of auto smog-control devices, the suit asks redress for damages caused in the state by auto exhaust pollution. The amount is unspecified, but state officials say it could run into billions of dollars.

Most companies that have been named as defendants in environmental suits are reluctant to discuss the details of the cases. But a spokesman for U.S. Steel, which has been involved in a number of cases, says, "We take all these suits very seriously."

Many of those filing pollution suits say they are doing so because they feel antipollution laws and those who enforce them simply don't do an adequate job. "If there's a theme to all these suits, it's an attempt to circumvent relatively ineffective regulatory agencies," says Joseph L. Sax, a University of Michigan law professor who is writing a book on environmental law.

In Illinois, under a controversial new law, state Attorney General William J. Scott can bring civil action on his own against any polluter, whether or not the matter is being considered by any other state agency.

"We saw that all the discourse relating to pollution was between industry and the boards of enforcement, which were more or less dominated by industry-oriented people. The public, which was being polluted, was never part of the discussion," says Allen A. Freeman, deputy attorney general. "So we developed the concept that the attorney general is the lawyer for the people and therefore will bring lawsuits on behalf of the people in the field of pollution."

Among the results is a suit filed against the city of Chicago, charging the city with illegal dumping and open burning of refuse at O'Hare Field. The suit alleges the burning causes air pollution and hampers the visibility of airline pilots when landing. It asks a permanent injunction to prevent the city from continuing the dumping and burning.

The attorney general's office also has filed suit against 23 airlines that use the airport, seeking to force them to install new pollution-control devices on all aircraft engines. Other suits seek court-enforced timetables that would force U.S. Steel and Republic Steel to clean up discharges by their plants into Lake Michigan and the Calumet River. Mr. Scott also has filed an antitrust suit against the four major auto makers, alleging a conspiracy to avoid installation of smog-control devices on cars.

To be sure, the criminal laws pertaining to pollution aren't being entirely ignored. The Federal Government has filed criminal suits against Olin Corp., Procter & Gamble Co., International Harvester Co. and others, charging them with polluting Chicago-area waterways. New Jersey's Department of Health is pushing a program against industrial pollution that has resulted in 140 complaints and $140,000 in court-ordered penalties.

At this point, no one knows what the results of most civil pollution suits will be. "Very few of these matters are beyond the pleading stage," says James Jeans, a professor at the University of Missouri law school who is chairman of the American Trial Lawyers Association's environmental law committee.

But, emboldened by a few early successes, environmentalists feel they stand a good chance of expanding the established legal concepts of nuisance or negligence to cover environmental questions.

Such legal concepts traditionally have been tied closely to property rights. "Until recently you didn't have a damage claim except for physically demonstrable damage, such as if your cow died or your tree died" as a result of pollution, says Pittsburgh lawyer Seamans.

But now environmental litigants are seeking both to prove harm that is other than economic and to assert rights less tangible than property rights. This has raised a question known to lawyers as "standing" to sue—or the ability to prove that you have a right to bring your problems before a court.

In the past, if a man couldn't show any direct personal economic harm, courts tended to throw his claim out on grounds that he didn't have any standing in the case. But in recent years the courts have become far more liberal in their concept of standing, permitting persons to come before the court as members of the public alleging damage to the public interest.

One of the most important decisions on the question of standing occurred in what is probably, in terms of legal precedent, the most important environmental case argued so far—the battle of conservationists to prevent construction of a power plant at Storm King Mountain on the Hudson River in New York. The conservationists succeeded in getting standing in court to challenge the Federal Power Commission's approval of Consolidated Edison Co.'s plans to build the plant. (At the moment, after years of litigation, the case is back before the FPC under a Federal court order requiring the FPC to reconsider its approval of the plant, taking environmental factors into account.)

The Storm King case established that conservationists do have standing to argue against planned projects that they contend would harm the environment.

Once in court, of course, environmentalists still must prove damage to the environment is occurring. "The courts want something a little more definite than people coming in and saying the sky isn't as blue as it used to be, or that the fumes are aggravating their sinuses," says Roderick A. Cameron, executive director of the Environmental Defense Fund, a Long Island-based coalition of lawyers and scientists.

Still, he notes, "the courts are now a little more inclined to move in the direction of protecting these intangible interests."

One of the problems faced by environmental groups is coming up with the technical data to prove that pollution exists and to show its effects. To help with technical problems, Mr. Cameron's Environmental Defense Fund has put together a committee of about 300 scientists who are willing to lend their expertise and testify in environmental cases. "We're acquiring a stable of expert witnesses," says Mr. Cameron, who, at 31, is himself both a lawyer and an engineer.

The fund also plans a program of mock trials at universities, using law students as lawyers and faculty members as expert witnesses so that all those involved will get some experience with pollution litigation. "Scientists are a little leery of getting involved in litigation, but once they get their feet wet they're a little less so," Mr. Cameron says.

—JOHN V. CONTI

1970